P. A. Bokhan et al.

**Laser Isotope Separation
in Atomic Vapor**

Related Titles

T. G. Brown, K. Creath, H. Kogelnik, M. A. Kriss,
J. Schmit, M. J. Weber (Eds.)

The Optics Encyclopedia

**Basic Foundations and Practical Applications.
5 Volumes**

2004. ISBN 3-527-40320-5

R. B. Firestone

Table of Isotopes

1999 Update with CD-ROM

1999. ISBN 0-471-35633-6

C. E. Little

Metal Vapour Lasers

1999. ISBN 0-471-97387-4

P. A. Bokhan, V. V. Buchanov, N. V. Fateev,
M. M. Kalugin, M. A. Kazaryan, A. M. Prokhorov,
D. E. Zakrevskiĭ

Laser Isotope Separation in Atomic Vapor

WILEY-VCH Verlag GmbH & Co. KGaA

The Authors

Petr A. Bokhan
Institute of Semiconductor Physics
Russian Academy of Sciences
Novosibirsk
Russia

Vladimir V. Buchanov
Institute of Chemical Physics
Russian Academy of Sciences
Moscow
Russia

Nikolai V. Fateev
Institute of Semiconductor Physics
Russian Academy of Sciences
Novosibirsk
Russia

Mikhail M. Kalugin
Scanning Laser Company
St. Petersburg
Russia

Mishik A. Kazaryan
P. N. Lebedev Physical Institute
Russian Academy of Sciences
Moscow
Russia

Alexander M. Prokhorov (1916–2002)

Dimitrij E. Zakrevskĭi
Institute of Semiconductor Physics
Russian Academy of Sciences
Novosibirsk
Russia

Cover picture   Based on an illustration from The Optics Encyclopedia, Vol. 2, p. 1021

■ All books published by Wiley-VCH are carefully produced. Nevertheless, authors, editors, and publisher do not warrant the information contained in these books, including this book, to be free of errors. Readers are adviced to keep in mind that statements, data, illustrations, procedural details or other items may inadvertently be inaccurate.

Library of Congress Card No: applied for

British Library Cataloging-in-Publication Data:
A catalogue record for this book is available from the British Library

Bibliographic information published by Die Deutsche Bibliothek
Die Deutsche Bibliothek lists this publication in the Deutsche Nationalbibliografie; detailed bibliographic data is available in the Internet at http://dnb.ddb.de

© 2006 WILEY-VCH Verlag GmbH & Co. KGaA, Weinheim

All rights reserved (including those of translation into otherlanguages). No part of this book may be reproduced in any form — by photoprinting, microfilm, or any other means — nor transmitted or translated into machine language without written permission from the publishers. Registered names, trademarks, etc. used in this book, even when not specifically marked as such, are not to be considered unprotected by law.

Typesetting   Hilmar Schlegel, Berlin
Printing   betz-druck GmbH, Darmstadt
Bookbinding   J. Schäffer GmbH, Grünstadt

Printed in the Federal Republic of Germany
Printed on acid-free paper

ISBN-13: 978-3-527-40621-0
ISBN-10: 3-527-40621-2

Foreword

Presently, there are many ways for isotope separation: gas diffusion, centrifugal, physicochemical, electromagnetic, optical, laser, etc. Each of them has proved to be suitable for separating specific classes of compounds in various aggregate states (gas, liquid, solid, or plasma) with various structures (atoms, two-atomic and polyatomic molecules, complex compounds, clusters) [1].

Optical and laser methods occupy a special place among other methods of isotope separation due to their high selectivity. In spite of the fact that first successful experiments on isotope separation were performed under lamp pumping [2–4], the modern development of optical methods is mainly based on employment of lasers. Presently, the possibilities of laser technique allow one to separate almost all chemical elements.

The fundamentals of laser isotope separation (LIS), including laser separation in atomic vapors, were discovered in the former USSR [5–7]. The theoretical and experimental study of physical processes responsible for LIS efficiency works on creating experimental laser complexes and systems for isotope separation [8–20] were also carried out in USSR. Nevertheless, the first promising experimental results on LIS in atomic vapors and considerable quantities of required isotopes were obtained in USA [20–22]. At that time this fact was related to the absence of a state program on the development of laser methods for isotope separation in atomic vapors. This branch was only developed in academic institutions.

At the same time, just after obtaining the promising experimental results at the Lawrence Livermore National Lab (USA) and many other countries large-scale research was started to create pilot and then semi-industrial laser complexes, first of all for uranium enrichment by ^{235}U isotope [20–30]. The real laser complex for isotope separation was based on a gas-discharge self-heating copper-vapor laser elaborated somewhat earlier [31–34]. It was this visible-range laser with unique characteristics and high repetition frequency that provided progress in laser isotope separation in atomic vapors.

Since the invention of a self-heating copper-vapor laser, progress in the development of such lasers was mainly determined by the efforts of native scientific groups (at P. N. Lebedev Physical Institute, Institute of Atmospheric Optics SB

Laser Isotope Separation in Atomic Vapor. P. A. Bokhan, V. V. Buchanov, N. V. Fateev,
M. M. Kalugin, M. A. Kazaryan, A. M. Prokhorov, D. E. Zakrevskiĭ
Copyright © 2006 WILEY-VCH Verlag GmbH & Co. KGaA, Weinheim
ISBN: 3-527-40621-2

RAN, NPO "Astrofizika" and Institute of High Temperatures [31, 32, 34]). Nevertheless, the strong government financing of such investigations in the USA (about $2 billion) has powered their leadership since the middle 1980s in the development of atomic vapor laser isotope separation (AVLIS). The term AVLIS has become a standard in scientific literature [35].

From 1972 to 1999, intense scientific investigations and pilot developments were carried out at the Lawrence Livermore National Laboratory, which resulted in the creation of a unique complex for the enrichment of uranium by ^{235}U isotope (used in nuclear power engineering), military plutonium by ^{239}Pu, gadolinium by ^{157}Gd, zirconium by ^{91}Zr, ytterbium by ^{168}Yb, and so forth. The production facilities helped to obtain hundreds of kilograms of required products. The complex of copper-vapor lasers developed, had a total average generation power of 72 kW, while the complex of tunable dye lasers had a power of 24 kW [20].

Similar AVLIS programs were started in France, Japan, and Israel [35].

A great interest in the practical mastering of laser isotope separation was also visible in Russia, Great Britain, China, India, Korea, and other countries. These investigations are partially reviewed in [1].

The qualitative evolution of a laser technique became noticeable in the last decade [36–56], which raises hope of a technological breakthrough in the production of industrial installations for isotope selection. It is significant that this method is universal for separating various isotopes, and can be applied to the separation of the required product from natural ores and industrial wastes with a high degree of purity.

Nevertheless, the works mentioned above were mainly carried out within the limits of the AVLIS method. Recent investigations carried out mainly in Russian scientific centers [44, 46, 49, 50, 54–56] show that there is a possibility of developing qualitatively new approaches to the problem of laser isotope separation in atomic vapors, which may result in the development of more efficient methods for isotope separation with accelerated accumulation of the required product.

The consideration presented below is devoted to the description of modern methods for isotope separation based on multiphoton coherent interactions and fast chemical reactions with selectively excited atoms. The results of the theoretical and experimental investigations of scientifically and practically important elements (Pb, Zn, Rb, B, Si), which can be used in fundamental investigations, in the development of quantum computers, and in microelectronic, atomic, and biomedical technologies, are generalized.

The choice of the considered elements was defined, on the one hand, by needs of science and technique, and on the other hand, by potential possibilities of modern laser techniques and competitive capabilities of laser methods as compared to other methods.

Attention is also paid to physicochemical aspects of isotope separation, the state and the evolution of laser technique.

Preface

Wide employment of isotopes in such fields as atomic and thermonuclear power, fundamental science, medicine, biology, isotopic geochronology, Mössbauer spectroscopy, agriculture, activation analysis, ecology, and production of new materials attracts increasing interest in the development of new highly efficient methods for isotope separation.

Modern development of optical spectroscopy, in particular, laser spectroscopy, makes it possible to obtain exhaustive information about the structures and shifts of spectral lines caused by isotopic effects. Recent progress in laser physics, methods of laser frequency tuning, control, and stabilization turns laser sources from laboratory devices to industrial installations. Laser methods for isotope separation have become easier to employ and new possibilities for obtaining isotopically modified and chemically pure substances have been opened. A unique possibility has arisen of not only separating isotopes of various atoms, but also separating isomers and isobars. This is important for mastering the industrial laser isotope separation and for further progress in fundamental investigations including the diagnostic problem of synthesizing new superheavy elements.

Laser isotope separation methods were developed in many countries in the framework of wide programs, first of all in the USA, France, and Japan. Most of the works were devoted to the method of selective photoionization, which was termed AVLIS (atomic vapor laser isotope separation) in these programs. Presently, it is necessary to develop more efficient methods for isotope separation anticipating their competitive ability in economy and ecology. In our opinion it has become possible, first of all, due to the development of laser spectroscopy and laser technique, investigations performed in the field of coherent interaction between radiation and atoms, in particular, the two-photon coherent effects, the nonlinear parametric processes, etc.

One more important feature of the development of modern methods is a great success achieved in studying single-photon and multiphoton light-induced chemical reactions with high rate constants.

In this book we present well-known investigations described in numerous publications that were performed by a conventional AVLIS scheme. The aim of this

Laser Isotope Separation in Atomic Vapor. P. A. Bokhan, V. V. Buchanov, N. V. Fateev,
M. M. Kalugin, M. A. Kazaryan, A. M. Prokhorov, D. E. Zakrevskiĭ
Copyright © 2006 WILEY-VCH Verlag GmbH & Co. KGaA, Weinheim
ISBN: 3-527-40621-2

book is to give a general description of the problem of laser isotope separation in atomic vapors. Attention is mainly paid to the development of the photochemical method of isotope separation, which has economical prospects for large-scale industrial production.

The Authors

The Authors of this Book

Petr A. Bokhan

Vladimir V. Buchanov

Nikolai V. Fateev

Mikhail M. Kalugin

Mishik A. Kazaryan

Dimitrij E. Zakrevskiĭ

Alexander M. Prokhorov
(1916–2002)

Contents

1	**Laser Isotope Separation in Atomic Vapors**	*1*
1.1	Introduction *1*	
1.2	Brief Description of the AVLIS Process as Applied to Uranium *3*	
1.3	General Description of the AVLIS Process *4*	
1.4	Theoretical Description of the AVLIS Process *6*	
1.4.1	Theoretical Description of the Method for Incoherent Interaction Between Radiation and Atoms *7*	
1.4.2	Features of Coherent Two-Photon Excitation *9*	
1.4.3	Evaporation of Separated Material, Collimation of an Atomic Beam, and Ion Extraction *10*	
1.5	Photochemical Laser Isotope Separation in Atomic Vapors *13*	
1.6	Other Methods of Isotope Separation *15*	
2	**Laser Technique for Isotope Separation**	*17*
2.1	Introduction *17*	
2.2	General Requirements for a Laser System in the AVLIS Process *18*	
2.3	Laser Complex *21*	
2.3.1	Pumping Lasers *21*	
2.3.2	Tunable Lasers *25*	
2.4	Complexes for Laser Isotope Separation *26*	
3	**Chemical Reactions of Atoms in Excited States**	*39*
3.1	General View of Photochemical Reactions *39*	
3.2	Experimental Study of Photochemical Reactions Between Atoms and Molecules *42*	
3.3	Collisional Quenching of Excited Atomic States by Molecules *46*	
3.4	Resonance Transfer of Excitation in Collisions *48*	
3.5	Collisional Processes with Rydberg Atoms *51*	
3.6	Isotope Exchange Reactions *55*	
3.7	Radical Reactions in Collisions *57*	

Laser Isotope Separation in Atomic Vapor. P. A. Bokhan, V. V. Buchanov, N. V. Fateev,
M. M. Kalugin, M. A. Kazaryan, A. M. Prokhorov, D. E. Zakrevskiĭ
Copyright © 2006 WILEY-VCH Verlag GmbH & Co. KGaA, Weinheim
ISBN: 3-527-40621-2

4	**Isotope Separation by Single-Photon Isotope-Selective Excitation of Atom** *59*
4.1	Description of the Method *59*
4.2	Mathematical Model of the Method *61*
4.3	Calculation Results on Isotope-Selective Excitation of Zinc Atoms *66*
4.3.1	Transversal Gas Circulation *67*
4.3.2	Longitudinal Gas Circulation *70*
4.4	Output Parameters Versus the Detuning of Radiation Frequency *71*
4.5	Influence of the Radiation Line Profile on Output Characteristics of the Separation Process *74*
4.6	Experiments on Laser Separation of Zn Isotopes by the Photochemical Method *78*
4.7	Experiments on Laser Separation of Rubidium Isotopes by the Photochemical Method *85*
5	**Coherent Isotope-Selective Two-Photon Excitation of Atoms** *91*
5.1	Brief Description of Two-Photon Excitation and the Mathematical Model *91*
5.2	Two-Photon Excitation of Led Atoms *93*
5.3	Two-Photon Excitation of Boron and Silica Atoms *95*
5.4	Photochemical Separation of Zinc Isotopes by Means of the Two-Photon Excitation *101*
5.4.1	Description of the Method *101*
5.4.2	Polarization of Radiation *103*
5.4.3	Mathematical Model of Cascade Superluminescence *105*
5.4.4	Calculation Results *108*
5.4.5	Experimental Results *111*
5.5	Zinc Isotope Separation by Evaporating Material from Chamber Walls *115*
5.5.1	Problem Statement *115*
5.5.2	Physical Analysis *118*
5.5.3	Calculation Results and Their Analysis *124*
5.5.4	Influence of Diffusion Processes on the Selectivity of Isotope Separation *127*
6	**Prospects for Industrial Isotope Production by Methods of Laser Isotope Separation** *131*
6.1	Microelectronics and Optoelectronics *133*
6.2	Nuclear Fuel Cycle *135*
6.3	Medicine and Biology *138*

7	**Appendix A: Mathematical Description of the Processes Based on Kinetic Equations** *139*	

8	**Appendix B: Operation Features of Copper-Vapor Laser Complexes** *141*	
8.1	Specificity of Creating the Complexes of Copper-Vapor Lasers *141*	
8.1.1	Specificity of Measuring Laser Radiation Parameters in CVL Complexes *147*	

9	**Appendix C: Physical and Technical Problems of Increasing the Power of Copper-Vapor Lasers** *151*	

10	**Appendix D: Neutron Transmutation Doping of Silica** *167*	

11	**Appendix E: Employment of Boron Isotopes in Microelectronics** *171*	

12	**Appendix F: Employment of Boron in Nuclear Fuel Cycle Equipment** *173*	

References *177*

Subject Index *185*

1
Laser Isotope Separation in Atomic Vapors

1.1
Introduction

For the last 30 years, the investigations and developments aimed at creation of new highly efficient, alternative methods for separating isotopes of chemical elements were actively performed in some countries and attempts were made for their industrial employment. Among those methods, isotope separation with the help of laser radiation is of particular interest. The methods of atomic (AVLIS process—photoionization method) and molecular (MLIS process) isotope separation are well known and developed.

Since the early 1980s, the further development of AVLIS process was additionally stimulated and brought up to a qualitatively new level due to achievements in laser spectroscopy, and plasma physics and technique. The acceleration in this period is mainly explained by the fact that in the USA the photoionization method of isotope separation was chosen after thorough study and comparison of the above-mentioned methods from the point of view of the highest accessible efficiency for uranium isotope separation.

Throughout 1980s, Lawrence Livermore National Laboratory (USA) in cooperation with some large companies extensively developed, updated, and demonstrated the elementary basis of AVLIS. Up to the late 1980s, the noticeable success in all directions of the program was achieved. In 1989, it became possible to start with specifying the date of the development of AVLIS technology for full-scale industrial production.

On January 1990, the USA Department of Energy presented to Congress a detailed program on AVLIS technology that included the order, conditions, and the date of demonstrating equipment in operating modes as well as the date of assignment of the developed technology and of setting into operation a powerful optical factory for uranium laser separation and concentration. This program was

Laser Isotope Separation in Atomic Vapor. P. A. Bokhan, V. V. Buchanov, N. V. Fateev,
M. M. Kalugin, M. A. Kazaryan, A. M. Prokhorov, D. E. Zakrevskiĭ
Copyright © 2006 WILEY-VCH Verlag GmbH & Co. KGaA, Weinheim
ISBN: 3-527-40621-2

approved by the Congress, and creation of new experimental demonstration equipment, a large-scale prototype of optical mill, started in Livermore.

Specialists of Lawrence Livermore National Lab had to create, demonstrate, and bring the new technology into commercial practice. The period from 1990 to 1995 was devoted to intensive preparation and carrying out a final series of demonstration tests of equipment, aimed at specifying the performance data and cost of the AVLIS process as a whole, which was needed for substantiating the full-scale production. The final strategic goal of the AVLIS program was to reduce the cost of uranium concentration via realizing new competitive technology.

During 1992–1993, an important development stage was accomplished and new-generation equipment was created. This equipment, including high-power laser systems and industrial-scale separators capable of providing the required efficiency of uranium concentration and production volumes, was subjected to complex demonstration tests. Some economical and technical problems concerning industrial production were eliminated in the course of the tests. The Department of Energy has thus come to a decision to build an optical mill.

For more than 10 years, the annual financing amounted to $150 million, because the AVLIS process was considered the cheapest technology for uranium enrichment to the level (the ratio of ^{235}U and ^{238}U isotopes) sufficient for using it as fission material.

The progress in realizing the AVLIS process in the early 1990s made it possible for the USA to start a new program on developing and demonstrating a technological basis for plutonium enrichment via the AVLIS process as the basic component capable of obtaining isotopically reach plutonium for military purposes.

Nuclear power plants in Japan make the basis of national power engineering. Apparently, it is believed that the difficulties inherent in nuclear power engineering are less risky than the dependence on oil import. Thus, in 1990–1995 the research and developments of the new, distinct from classical (centrifugal and gas-diffusion separation), technology of uranium isotope laser separation in atomic vapors, had got an additional acceleration. The corresponding investments to the development of the AVLIS process amounted to hundreds of millions dollars. The Japan national program devoted to solving this problem is also termed AVLIS. Ideally, the new technology of uranium enrichment should be a technology of 21st century, that is, compact, highly automated, and the safest one. In view of Japan territorial specificity, the radiation safety is of particular importance. A load on environment and the probability of radiation and chemical damage of population by concentrator wastes should be minimized.

In Japan, the works concerning the AVLIS process started at University of Osaka and Japan Energy Research Institute. Presently, the most active center succeeded in investigations and development of AVLIS technology is Laser Atomic Separation Engineering Research Association of Japan (LASER-J) in Tokyo. It is a joint industrial–government organization founded in April 1987 by nine electrotechnical companies, Japan Atomic Power Company, Japan Nuclear Fuel Industries Company, and Central Research Institute of the Electric Power Industry.

Presently, financing of the AVLIS program is of the order of $154 million (two third of this sum is provided by government budget) for creating an industrial AVLIS prototype within the next 4–5 years.

The corresponding European program is called SILVA. The most important components for laser isotope separation (LIS) are developed in France (Jilas Alcatel) and England (Oxford Lasers). In recent years, high investigation and development activity in this field has also been observed in China (Shanghai University) and Israel (Beer Shiva University).

1.2
Brief Description of the AVLIS Process as Applied to Uranium

In the beginning, LIS ideas in physical and industrial aspects were developed in the framework of national AVLIS programs on uranium enrichment. Some key steps and stages of this process are published in [64–67].

Let us briefly describe the AVLIS process as applied to uranium.

It is known that uranium conversion to appropriate fission fuel necessitates the concentration of ^{235}U from 0.7 % (the natural contents of isotope in a mixture with ^{238}U) to 3 %. In AVLIS technology, atomic vapor that is a natural uranium isotope mixture is obtained by electron–vacuum evaporation in a special high-vacuum technological unit. Then vapor passes from the evaporation unit to flow former, where it is shaped to a required, for example, leaf-like form. Then it comes to an interaction zone with light beams produced by a dye laser system. The dye laser system is pumped by another high-power system of copper-vapor lasers. The ionization potential for uranium is $U_{ion} = 6$ eV; the isotopic shift of uranium levels amounts to 0.08 cm^{-1}. Then uranium atomic vapor is photoionized in the interaction zone via a three-step excitation process. The spectral width of radiation, the operating wavelengths λ_1, λ_2, λ_3, and the average power of dye laser beams are matched in such a way that after absorbing three sequential quanta of light at specified uranium atom transitions, only the required isotopes would be excited in cascade and photoionization processes. The operating levels of uranium atom are schematically shown in Fig. 1.1. Quanta of energy for each transition are near 2 eV. The choice of particular levels is a rather complicated spectroscopic problem. Uranium possesses a branched system of atomic levels and has many valence electrons. The wavelengths of possible laser sources are also shown in Fig. 1.1 [27].

When a copper-vapor laser pumps dye lasers, all three wavelengths of the dye lasers are chosen in the red spectral range because a uranium atom has allowed transitions between levels suitable for efficient cascade excitation in this range (590–600 nm). In the three-step scheme, an atom at the upper autoionization

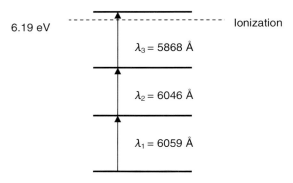

Fig. 1.1 Photoionization of a uranium atom.

level decays into electrons and ions, that is, it is ionized. High selectivity of the preferred ionization for the desired isotope is obtained at all cascade-excitation steps due to fine matching between the frequencies of tunable lasers and the energy levels of the uranium atom. ^{235}U isotopes ionized in this way are extracted from the interaction zone by an electric field and are directed to collector plates where they are neutralized and condensed as the final product of separation.

1.3
General Description of the AVLIS Process

The method of isotope separation considered above has a long history (more than 30 years). The list of papers devoted to the description of this method would make dozens of pages. In this section, we consider only principal features of the AVLIS method. A more detailed description can be found in [20]. A detailed consideration of the method applied, in particular, to ytterbium isotope separation is given in review [19].

The AVLIS method is based on multistep isotope-selective ionization of atom. Generally speaking, the upper excited level can be ionized in different ways. Conventionally, a laser radiation is used. Isotope-selective excitation of atom became possible by means of a frequency-tunable laser radiation with the spectral line width less than the specific frequency differences between the absorption lines of isotopes under consideration.

One important problem with the method considered is the choice of a photoionization scheme. It may be a two-level, three-level (see Fig. 1.2), or four-level scheme. The choice is mainly determined by the laser source. Most of the atoms have such a structure of energy levels that the first-transition radiation (with the wavelength λ_1 in Fig. 1.2) should lie in the visible or UV spectral range. For the second transition, the IR radiation may be used. The corresponding lifetimes are

1.3 General Description of the AVLIS Process

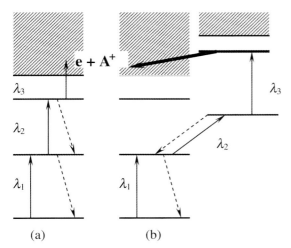

Fig. 1.2 Stepwise ionization: (a) direct photoionization; (b) ionization via the autoionization state (heavy line). The dotted lines designate the radiation decay channels for levels; the thick arrow refers to the decay of the autoionization state into electrons and ions.

usually not very long and lie in the range 3–200 ns (we do not consider excitation of metastable or Rydberg levels).

To avoid losses to spontaneous emission one should employ pulse-periodic sources of laser radiation with the pulse duration lying in the range mentioned.

The characteristic excitation time for a level is determined by the formula $\tau = (\sigma\Phi)^{-1}$, where σ is the absorption cross-section and Φ is the quantum flux. The energy of the pulse per unit area is $W = h\nu\Phi\tau = h\nu/\sigma$. The magnitude of σ is at least 10^{-15} cm^{-3} even in the case of Doppler broadening. For the characteristic quantum energy of 1 eV we have $W \simeq 160$ µJ cm^{-2}. At the pulse repetition frequency of 10 kHz (the typical frequency for pumping pulses) we obtain that the required power is in the range 1–2 W cm^{-2}. The radiation power of modern laser sources noticeably exceeds this value in UV, visible, and IR spectral ranges.

Energy levels are excited by moderate energy, whereas direct photoionization necessitates energy a few orders higher. This is explained by the small cross-section of photoionization, which is 10^{-17}–10^{-18} cm^2. Under the above-mentioned conditions, the average power of the ionizing radiation is estimated as 100–1000 W. The low cross-section of direct photoionization results in a weak absorption of the ionizing radiation. Hence, the optical density is a few orders lower than the corresponding value for the radiation exciting atomic levels. A serious problem of matching optical paths arises at various steps of photoionization. The photoionizing process will mainly determine the energetics of isotope selection.

Ionization via the decay of autoionization levels can be used instead of direct photoionization (see Fig. 1.2b). Such levels lie higher than the first ionization potential. Emission lines from autoionization states are observed in most of chemical elements. Presently, identification of autoionization levels, and refinement of their positions and lifetimes are a subject of intensive investigations. The cross-section of such levels may amount to 10^{-15}–10^{-16} cm^2. Hence, the problems mentioned above are to a certain degree tempered. Search for appropriate autoionization levels is one of the most important problems in spectroscopy. By tuning the frequency of a tunable laser one can find from the electron yield due to ionization or luminescence processes the efficiency of exciting autoionization levels. The radiation lifetime of autoionization levels is usually a few orders longer than the time of atom autoionization decay into electrons and ions. Hence, we can neglect the losses related to spontaneous emission from such levels.

Conventionally, the isotopic structure with hyperfine line splitting is rather dense and overlapped with Doppler broadening. In this case, the Zeeman effect can be used by applying a magnetic field in order to separate the lines by a frequency shift sufficient for selective excitation of atoms. In particular, this method is recommended for palladium isotopes [68].

One more method of photoionization is to excite long-living Rydberg states lying close to the ionization threshold. A relatively high ionization cross-section and proximity to the ionization threshold make employment of powerful infrared lasers (CO_2, CO, etc.) probable.

Under an electrostatic field applied, Rydberg levels shift. If then the electric field is switched off, some high levels may become higher than the ionization threshold and the atom may decay into electrons and ions. A drawback of such ionization via Rydberg states is a high sensitivity of the state positions to weak induced external electric fields, which smears the levels and hinders their excitation.

1.4
Theoretical Description of the AVLIS Process

We have considered above some basic methods of photoionization. Two physically different cases are observed in actual investigations of laser isotope separation. In the first case, incoherent interaction between radiation and atoms occurs, whereas in the second case this interaction is principally of coherent character. In the incoherent interaction model, radiation is considered as a flux of photons, which excite or de-excite atomic levels. The radiation itself is described by transport equations for a flux of particles. This approximation is often used in laser kinetic calculations. Equations for incoherent approximation are considered in more details in many papers, see, for example, review [19].

Coherent interaction of radiation with atoms is considered by means of a density matrix and the radiation itself is described by the equation for the electric field intensity that follows from the Maxwell equations.

Incoherent interaction between radiation and atoms occurs if the radiation coherence time or time of level transversal relaxation is much shorter than the characteristic time of variations of the atomic level population (actually it is comparable to the radiation pulse duration). In the opposite case, the interaction is of coherent character.

If the isotopic frequency shifts lie in the gigahertz range, then the selectivity of atom excitation is provided by the radiation with a considerably broad spectral width of line (0.5–1 GHz). This value is in accordance with the estimate $\Delta \nu \tau_{coh} \simeq 1$ (here $\Delta \nu$ is the effective line width and τ_{coh} is the characteristic coherence time) corresponding to the coherence time $\tau_{coh} \simeq 1$–2 ns. An actual pulse duration of the frequency-tunable visible or UV radiation lies in the range 5–50 ns. Hence, the coherence time is much shorter than the duration of the radiation pulse and the process of isotope separation can be described in the approximation of incoherent interaction. If the isotopic shifts are less than 1 GHz, then the incoherent approximation is rarely applicable.

1.4.1
Theoretical Description of the Method for Incoherent Interaction Between Radiation and Atoms

We have considered some basic photoionization methods. However, physics of level excitation may differ for radiation pulses with different durations and line widths. At early investigation stages, mainly a broadband radiation was used. The spectral width of radiation exceeded the inverse pulse duration time or inverse radiation lifetime. Under these conditions, the analysis may be performed in the incoherent approximation of the interaction between radiation and atoms.

The method of isotope separation considered in the case of incoherent interaction can be described in the framework of the kinetic model, which is widely used for the analysis in gas laser physics. We will assume Doppler distribution of atoms over velocities. Let us first consider the point model without taking into account spatial variations of parameters. For better perception we also assume that the spectral width of the atomic absorption line is far less than that of the radiation line and of Doppler profile. Later, we will make corrections for the case where such assumptions are not applicable.

Under the conditions mentioned, we may assume that binary impacts between atoms and photons corresponding to frequency ν occur during the radiation pulse. For such cases, we will employ the density of atoms and photons per unit frequency interval and for other cases the total particle concentration will be assumed. As is shown in Appendix A, the kinetic equations can be obtained automatically. It suffice to indicate all the processes in the form of chemical reactions with the corresponding rate constants. Reactions given below are the same for all isotopes.

Excitation and depopulation of atomic levels by radiation is described by the following reactions:

$$N_i + h\nu_i \Rightarrow N_{i+1}; \quad \alpha = ck_i(g_{i+1}/g_i)$$
$$N_{i+1} + h\nu_i \Rightarrow N_i + h\nu_i + h\nu_i; \quad \alpha = ck_i$$

Here N_i is the atom at the ith level; $h\nu_i$ refers to the radiation in a transfer from the $(i+1)$st to the ith level. To the right of the reactions, the corresponding rate constants are given, where c is the speed of light; g_i is the statistical weight of the ith level; and $k_i = A_{i+1}\lambda_i^2/(8\pi)$ (λ_i is the wavelength and A_{i+1} is the Einstein factor for the transition considered).

The photoionization reaction looks like

$$N_{up} + h\nu_{ion} \Rightarrow N^+ + e \tag{1.1}$$

where N_{up} is the upper level in the stepwise scheme of photoionization; $h\nu_{ion}$ refers to the ionizing radiation; and N^+ and e denote ions and electrons, respectively. The rate constant depends on the details of atom excitation to the autoionization state. It is worth noting that the corresponding experimental data seem to be more reliable than the theoretical formulae available.

If a spontaneous decay of level occurs at the preceding level in the scheme of stepwise ionization, then we can write

$$N_{i+1} \Rightarrow N_i; \quad \alpha = A_{i+1} \tag{1.2}$$

and if the decay occurs at the levels that are not taken into account in the model, then we can write

$$N_{i+1} \Rightarrow \quad ; \quad \alpha = 1/\tau_{i+1} - A_{i+1} \tag{1.3}$$

Here, τ_{i+1} is the total radiation lifetime of the level.

In certain cases, one should also take into account the reaction of energy transfer in impacts from the atom of one isotope to the atom of another isotope:

$$N_i^* + N_j \Rightarrow N_j^* + N_i \tag{1.4}$$

where the indices refer to different isotopes.

The concentration of atoms can be limited above due to the charge exchange reaction for ions:

$$N_i^+ + N_j \Rightarrow N_j^+ + N_i \tag{1.5}$$

The rate constants for the reactions depend on the distribution function of particles over velocities. Calculation of rate constants for a particle flux is a complicated problem and beyond the scope of this book.

The impact of heavy particles will be considered in more details in Chapter 3.

The reactions considered describe mainly a conventional point model of the process of pulse radiation. Nevertheless, this model neglects the possibility of superluminescence from one of the considered levels to the levels that are not directly excited by radiation in the process of stepwise photoionization. The latter factor is important if the upper state comprises Rydberg states. The cross-sections of induced transitions between them are so high that a superluminescence occurs at the density much lower than a typical operating concentration of atoms. A mathematical model of cascade superluminescence is presented in Section 5.4.3.

In lifting restrictions on the widths of the atomic absorption line and radiation line, one should make certain corrections. The excitation and de-excitation rate constants are multiplied by the parameters depending on the line profiles.

Let the profile of the atom absorption line be given by the normalized function $L(\nu)$. Then, these parameters are

$$\alpha_N(\nu) = \int_0^\infty \frac{n_{\text{ph}}(\nu' - \nu)}{n_{\text{ph}}(\nu)} L(\nu') d\nu', \qquad \alpha_s(\nu) = \int_0^\infty \frac{N_i(\nu' - \nu)}{N_i(\nu)} L(\nu') d\nu'$$

where n_{ph} is the photon density at the corresponding transition.

In the equations for level populations, the rate constants of excitation and de-excitation by radiation are multiplied by α_N. In the equations for photon density, the rate constants are multiplied by α_s.

In a one-dimensional model, variables depend on the coordinate along the laser beam. The only change to be made in writing the equations is to substitute the ordinary derivative by a substantial one in the photon density equation:

$$d/dt = (1/c)\partial/\partial z + \partial/\partial t \tag{1.6}$$

where z is the coordinate along the laser beam.

1.4.2
Features of Coherent Two-Photon Excitation

Development of the laser technique resulted in an extremely small width of the radiation line limited only by the pulse duration. A typical pulse duration of most powerful sources of tunable radiation is in the range 5–50 ns, which is close to the characteristic time of level radiation decay. In this case, the coherent model of the interaction between radiation and atoms should be used. The model is based on the Liouville equation for the density matrix of a multilevel system:

$$i\hbar(\partial/\partial t + \hat{\Gamma})\hat{\rho} = [\hat{H}_0 + \hat{V}, \hat{\rho}] \tag{1.7}$$

where $\hat{\Gamma}$ is the operator describing system relaxation; $\hat{H}_0$ is the Hamiltonian of the undisturbed system; and $\hat{V}$ is the operator responsible for the interaction with the electric field of radiation.

A detailed writing of Equation (1.7) will be given in Chapter 5.

In the coherent approximation, excitation of the third upper level is not considered as subsequent excitation via an intermediate level. Interaction of a multilevel system with radiation occurs as a single process. Under optimal conditions, it is possible to obtain almost 100 % excitation of the upper level at an extremely narrow exciting line. This is the advantage of the coherent excitation as compared to a conventional scheme, where the absorption lines can undergo field broadening by the radiation and by decay radiation of levels. In the coherent case, the frequency tuning may help reduce an influence of an intermediate level on the excitation efficiency and on the width of the absorption line. Coherent two-photon excitation will be considered in particular examples in Chapter 5.

1.4.3
Evaporation of Separated Material, Collimation of an Atomic Beam, and Ion Extraction

In addition to the consideration of photoionization problems, the AVLIS method requires solving some other important problems. Generally, isotope separation includes the following stages: evaporation of material; producing a collimated atomic beam; ion extraction and acquisition of certain isotopes at collectors; photoionization of atoms; and gathering and extracting atoms not subjected to photoionization (excavation system and gas evacuation from the working chamber), see Fig. 1.3.

Volatile elements can be obtained with a crucible-type vapor source, where material is evaporated in a closed volume and gets out through slits. At high pressure, a gas-dynamic jet can be formed in the evaporation chamber. In the latter case,

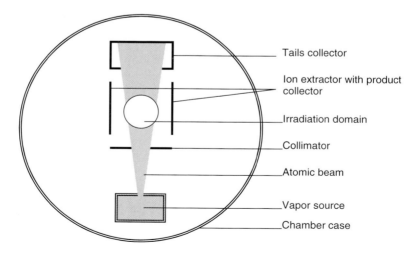

Fig. 1.3 Schematic diagram of installation in a cross-sectional view.

clusters capable of scattering and absorbing radiation may probably be formed. At low pressure, collision-free gas flow is used. The advantage of the crucible evaporation method is simple construction and low voltages applied across heating elements.

Electron-beam evaporator may be an alternative vapor source. Its principle of operation is as follows. An electron beam hits the surface of the material to be evaporated and heats it to a high temperature, so that gas-dynamic vapor flow is produced. While expanding, the vapor is cooled, which results in low temperatures in the irradiation zone ($\simeq$ 200 K). Such a temperature favors the reduced Doppler broadening and lower concentration of low metastable atomic states. High voltage, high coefficient of reflecting electrons from a surface, and undesirable vapor ionization are drawbacks of the method.

In view of hyperfine splitting, the isotope line structure is partially or completely overlapped by the Doppler profile in most of the atoms. Collimation of an atomic beam is used to reduce the Doppler width. Across the vapor path, the construction is placed made of tubes or plates with slits. By this way, a rather narrow atomic beam is separated from the flux with comparatively small velocity deviation along the radiation beam. Atoms not passed through the slits are removed by the gas-evacuation system. It is a great drawback of the collimation system that only negligible part of the atomic beam passing from the evaporator is used in the isotope separation process. The remaining atoms produce an undesirable background, which reduces the selectivity of isotope separation because atoms may fall to the collectors intended for specific isotopes. High vacuum needed in this case ($10^{-6} - 10^{-7}$ Torr) requires employment of a cumbersome pumping system.

An extractor is used to draw out ions from the atomic flux. There are various methods for extracting ions. The most known one is electrostatic extraction. A voltage is applied across the electrodes of the extractor, which provides ion extraction at the cathode.

The photoionization results in producing quasi-neutral plasma. The time of electric field penetration into the plasma depends on the density of the latter. Under a voltage pulse, the plasma polarizes due to shift of electrons toward the anode. A thin (relative to the electrode separation) layer of ions outcrops at the cathode: $d = (U/2\pi en)^{1/2}$, where U is the potential difference and n is the plasma density. Under standard conditions ($n = 10^9$ cm^{-3} and $U = 100$ V) the estimate yields $d \simeq 0.3$ cm. All the potential difference is applied across this layer. Then it is depleted of ions due to their passing to the cathode. Hence, the cathode layer width increases and the boundary of the quasi-neutral plasma shifts to the anode. Efficient ion extraction occurs when the cathode layer width becomes close to the separation between the electrodes. It is also required that the transit time in the plasma was longer than the extraction time. As is shown in [69], in the plasma with density 10^{10} cm^{-3} the extraction time is $\simeq$ 100 μs at the separation of 2 cm between the electrodes, which is much longer than the time of the plasma passage through the electrode zone.

The situation becomes better if the voltage between the electrodes is constant and the plasma enters this domain permanently. Such a regime can be realized by means of photoionization in the domain behind the extraction zone along gas flow and at high repetition frequency of pulses, where the transit time is not longer than the time interval between pulses. The latter condition can be realized at conventional repetition frequencies ($\geq$ 10 kHz). In this case, the stationary layer of the cathode drop is established that expands up the flux. Estimates show that at acceptable distances between the electrodes the ion extraction can be efficient.

At a higher plasma density, the role of electron gas pressure becomes noticeable, which is higher than the pressure of ions because of the higher electron temperature [69]. Electron gas tends to expand and involve ions producing ambipolar flux to peripheral of the extraction zone. Plasma expansion occurs at the speed of the ion sound. The extraction time for the interelectrode separation of 2–3 cm is estimated as 20–60 µs, which also exceeds the characteristic transit time for the plasma. The plasma also expands in the longitudinal direction, which reduces the number of ions hitting the electrodes. In addition, ions motion toward flux increases the probability of undesirable charge exchange in impacts with neutral atoms.

The limiting plasma density is estimated to be 10^{10} cm^{-3} in the extraction processes described above.

An alternative extraction method is crossed electric and magnetic fields. It is known that the electric field in plasma accelerators penetrates quite dense quasineutral plasma. This is possible due to drastic fall in the mobility of magnetized electrons along the electric field direction (see Fig. 1.4). For electrons to be magnetized, their Larmor radius should be much shorter than the separation between the electrodes. In contrast, the Larmor radius for ions is much longer than this separation. Strongly magnetized electrons drift at the velocity cE/H along equipotential surfaces (see Fig. 1.4), thus forming the Hall current, whereas ions accelerate in an electric field not affected by magnetic forces. Since there is no electron drift in the direction of the electric field, the latter almost completely penetrates the plasma.

The principle of ion acceleration is quite clear; however, in practice, there are certain difficult problems: to shunt the Hall current; provide emission of electrons, compensating ions departure to the cathode; match extractor geometry with the construction of a separation chamber, etc.

According to calculations [69], in the Hall extractor the ion density may be as high as 10^{12} cm^{-3}.

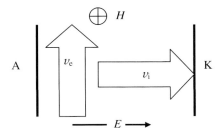

Fig. 1.4 Ion acceleration in crossed electric and magnetic fields (v is the drift velocity).

1.5
Photochemical Laser Isotope Separation in Atomic Vapors

Long-term intensive investigations of the laser photoionization method for isotope separation revealed the following drawbacks in industrial employment:

- large volumes of separation chambers needed because the density of the separated isotope is low;
- maintaining high vacuum in order to avoid a background density of atoms;
- complicated cumbersome optical systems needed for aligning and focusing beams with the optical length of dozens or hundreds of meters;
- a complicated system for ion collimation and extraction;
- the necessity of using gas fluxes at near-sound velocities, which results in that atoms are irradiated only once during the transit time.

These drawbacks are absent in the laser photochemical method of isotope separation, though it has its own negative features. However, historically the photoionization laser method for isotope separation was preferred. The photochemical method did not actually develop. Only in the last few years the interest to this method has tend to increase and the prospects were being determined.

Rapid evolution of photochemistry reveals a large number of photoinduced reactions. It seems that for most of the chemical elements, specific reagent can be found capable of binding excited atoms, thus forming a molecule. However, in a fundamental state an atom has no energy sufficient for breaking chemical binds and it does not chemically react with reagent. The molecules formed in this way may be separated from a gas mixture by known physicochemical methods.

The development of a laser technique made it possible to selectively excite any atomic level of a particular isotope. For laser isotope separation, long-living (metastable) states are preferred. This is explained by the following. It is obvious that the chemical interaction probability for an excited atom should exceed the probability of radiation decay of the level, that is, the condition

$$k^* N > 1/\tau \qquad (1.8)$$

should hold, where k^* is the rate constant of the reaction; N is the concentration of reagent molecules; and τ is the radiation lifetime. The molecule concentration is limited to the approximate value 10^{17} cm^{-3}. At higher concentrations, the undesirable impact broadening of lines becomes noticeable (see Chapter 3) as well as the isotope exchange given by the reaction

$$A_1 R + A_2 = A_2 R + A_1 \tag{1.9}$$

where A_1 and A_2 are two different isotopes and R is a molecular group. Hence, even at sufficiently high rate constant ($k^* = 10^{-10}$ cm^{-3}) the condition $\tau \gg 100$ ns should be met. Consequently, the excited states in which an atom reacts chemically are interesting if their radiation lifetime is not shorter than 1 μs.

Possible schemes for exciting long-living levels are shown in Fig. 1.5. The simplest one is single-photon excitation of a long-living level (see Fig. 1.5a). If the isotope structure of levels is smoothed by the Doppler profile, then the isotopic selectivity is provided by frequency detuning from the central radiation frequency (see Chapter 4).

One more scheme is based on two-photon excitation of a high level followed by a decay into a long-living state (see Fig. 1.5b). The decay may occur due to spontaneous transitions as well as result from the development of superluminescence from the upper level. For the intermediate level, a long-living one is preferable, because it can be populated due to spontaneous emission from the upper level. Coherent two-photon excitation of the upper level is desirable because of its higher selectivity. With oncoming radiation beams, it is possible to noticeably reduce the width of the Doppler profile under two-photon excitation. The scheme considered has been used for separating zinc isotopes (see Chapter 5).

Metastable states are often disposed lower than the resonance level (Pb, Ba, Cu, Au, Mn, Eu, etc.). In this case one can employ the scheme shown in Fig. 1.5c. Two-photon coherent excitation is also preferable (in addition to the reasons mentioned above it helps in suppressing the superluminescence from the upper level

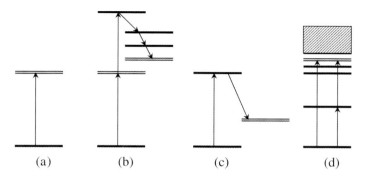

Fig. 1.5 Photoexcitation of long-living states (marked with a double line).

to a metastable level, which prevents the broadening of the absorption line and hindering the isotope selectivity during excitation).

Rydberg levels reveal high chemical activity due to a long distance between the valence electron and nucleus. Such levels are naturally long-living. Single- and two-photon schemes for exciting Rydberg states can be used (see Fig. 1.5d). A single-photon scheme was realized for separating rubidium isotopes (see Chapter 4).

In the photochemical method of isotope separation, the concentration of atoms is limited by the resonance transfer of excitation between atoms of different isotopes. This process certainly worsens the excitation selectivity. A cross-section of the excitation transfer in the dipole–dipole approximation is inversely proportional to the lifetime of the upper level [70]. The dipole momentum of a long-living state is not large and we may hope that the cross-section would be by a few orders lower as compared to typical conditions of the excitation transfer from a resonance level, in which case it is in the range 10^{-12}–10^{-13} cm^{-2} [70]. Estimating the cross-section of the excitation transfer by the value 10^{-13} cm^{-2} we obtain the above limit for the atomic concentration of the order of 10^{13} cm^{-3}. This value is three orders of magnitude higher than the limiting atom concentration acceptable in photoionization isotope separation.

The optical thickness in the photochemical method is much greater as compared to the photoionization method due to high concentration of atoms. The dimensions of the chamber are far smaller. A possibility arises to directly excite long-living levels, including metastable ones. Isotope-selective excitation of atoms becomes real at the frequency detuning from the central part of the absorption line equal to several Doppler widths (see Chapter 4). The velocity of gas flux reduces to 1–10 m s^{-1} and is provided by simple gas pumping through the chamber. Both the transversal and longitudinal gas flux are possible relative to the laser beam. These obvious advantages of the photochemical method for isotope separation stimulated intensive investigations of this approach. Investigations of isotope separation by the laser photochemical method on a large-scale installation in Russia were started only a few years ago; however, the results obtained show that this method can compete with the conventional photoionization method.

1.6
Other Methods of Isotope Separation

For separating small quantities of isotopes in atomic vapors some alternative methods were developed. Let us briefly discuss only two of them.

In [71], the method is described based on the effect of light pressure on atoms. The laser radiation crosses the atomic beam at a right angle. The radiation frequency is matched to excite certain isotopes. By absorbing light quantum, the atom acquires the momentum h/λ in the transversal direction with respect to the atomic beam. Spontaneous decay of a resonance level imparts a momentum of the same value, however, of an arbitrary direction to the atom. On average, the atom acquires a momentum Nh/λ, where N is the number of photon absorption events. Finally, such atoms escape atomic beam and isotopes are separated.

Another method is based on the known fact that particles with the magnetic momentum distinct from zero deviate from their trajectory in a nonuniform magnetic field. If an atom is excited to the state with the magnetic quantum number $m \neq 0$, then it acquires a magnetic momentum. It can be done by means of resonance radiation with circular polarization. It is important that if the level with $m \neq 0$ decays into metastable levels, then the latter acquire an average magnetic momentum distinct from zero. In a magnetic field, the upper level splits into sublevels with different m numbers. In this case, isotope-selective excitation simplifies, because the overlapping of the radiation lines of different isotopes can be avoided. In a nonuniform magnetic field, selectively excited atoms with nonzero magnetic momentum decline from a primary trajectory (see Fig. 1.6); the isotope composition in the atomic beam changes. In this way barium isotopes were separated [72].

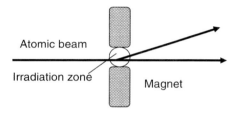

Fig. 1.6 Deviation of polarized atoms in a nonuniform magnetic field.

As far as LIS is concerned, the method of so-called photoinduced diffusion of atoms and molecules in the field of a monochromatic traveling wave [73–76] seems promising. The essence of the effect is that the velocity distribution for particles in excited or unexcited states becomes, due to the Doppler effect, asymmetric under the action of radiation with the frequency slightly different from that of atomic or molecular transition. In a mixture of buffer gas, such asymmetry results in an origin of the flux of absorbing particles directed along or opposite to the direction of radiation depending on the frequency detuning. By varying the experimental conditions, mainly the frequency detuning, one can spatially separate different isotopes. In [77], it is reported about approximately 70 % enrichment with respect to isotope ^{13}C in working with CH_3F molecule.

2
Laser Technique for Isotope Separation

2.1
Introduction

The fundamental principle of all optical methods for isotope separation is the precise action on a particular kind of atoms via selective absorption of narrow-band radiation. Then the excited or ionized atoms are extracted.

The idea of using optical radiation to selectively excite atoms or molecules of a particular isotope composition with the following photochemical separation arose just in the wake of the discovery of isotopes and the isotopic effect in atomic and molecular spectra. It seems that first successful experiments in this field were the selective action on certain mercury isotopes in a natural mixture by a 253.7 nm resonance mercury line, passed through a resonance absorbing filter, observations of a photochemical reaction between excited atoms with oxygen, and excitation of phosgene molecule by a 281.62 nm spectral line emitted by aluminum spark. There were also successive experiments with other elements [11]. Nevertheless, in all experiments unintentional coincidences between strong spontaneous emission lines and atomic or molecular absorption lines were used. The number of such coincidences is limited and narrow lines of spontaneous emission have low intensity; hence, the optical method of separation with nonlaser radiation sources could not be successfully applied despite of its advantages.

Lasers have markedly expanded the possibilities of optical methods for isotope separation. A laser source of optical radiation possesses rather useful features. Stimulated emission can be obtained at an arbitrary frequency in UV, visible, and IR spectral ranges. This provides operation with objects, which have absorption lines in various spectral ranges. High spatial coherence makes it possible to collimate radiation for irradiating lengthy volumes of material. Monochromaticity and high time coherence, high power, and intensity of radiation are sufficient for saturating absorbing transition and the small width of a spectral line pro-

Laser Isotope Separation in Atomic Vapor. P. A. Bokhan, V. V. Buchanov, N. V. Fateev,
M. M. Kalugin, M. A. Kazaryan, A. M. Prokhorov, D. E. Zakrevskiĭ
Copyright © 2006 WILEY-VCH Verlag GmbH & Co. KGaA, Weinheim
ISBN: 3-527-40621-2

vides excitation of one particular isotope. Employment of laser sources revealed the advantages of optical isotope separation [78]: high selectivity in the elementary separation act; possibility of separating specified isotopes; contact-free interaction; low-cost energy consumption; universality and mobility of the separation method.

In recent years a considerable progress has been made in laser physics and the methods of tuning, control, and frequency stabilization of laser radiation. This turns laser sources from scientific and laboratory devices to actual industrial installations, opens a wide variety of new applications, and simplifies the employment of LIS methods for obtaining isotopically and chemically pure materials.

It is obvious that the primary factor in the optical method of isotope separation is the action of tunable, narrow-band, frequency- and power-stable radiation with the wavelength matched (in resonance or at certain detuning) with a saturation line of a separated isotope. Laser sources capable of generating tunable coherent radiation are needed, with the required parameters in a prescribed spectral range. For LIS purposes, most interesting are dye lasers and, currently, tunable solid-state (in particular, titanium–sapphire) lasers. Such lasers require optical pumping, for which coherent radiation of a copper-vapor laser (CVL) is mainly used. In recent years an interest has been shown to rapidly developing solid-state devices. If the object under study requires UV radiation, then the systems for frequency doubling in nonlinear crystals are used.

2.2
General Requirements for a Laser System in the AVLIS Process

Let us now formulate general requirements for a laser system that follow from the mechanism of the AVLIS process as applied to uranium separation and briefly comment on them.

(1) According to the three-step scheme chosen for the cascade photoionization, the radiation of all three laser beams interacting with uranium atoms should reside in the red–orange spectral range. Radiation of all dye lasers should be narrow-band, tunable over wavelengths, and provide precise adjustment and active stabilization at a chosen wavelength.

As was mentioned above, the isotopic shift for uranium energy levels is approximately $\Delta\nu = 0.08$ cm^{-1} or 2.4 GHz. Radiation of the laser system should be precisely matched in such a way that a sequential absorption of all quanta from each of the chosen levels mainly occurred in the spectral bands corresponding to isotopes ^{235}U without overlapping with the bands of ^{238}U.

The required selectivity of tuning and operation in the working wavelength bands can be easily obtained with tunable dye lasers. There are active media (rhodamine, oxazine) for dye lasers efficiently operating in the red–orange range under the pumping by radiation in a green or yellow spectral range. With gas-discharge copper-vapor lasers as a source for pumping the latter requirement is completely satisfied.

(2) The laser system should operate in a pulse-periodic mode at a high repetition frequency of radiation, which may be from units to dozens of kilohertz.

The interaction between the laser radiation and atoms should run in the following way. Strongly synchronized laser pulses enter the interaction zone at the moment, when it is filled with unexcited uranium atoms that are mainly in the ground state. They are irradiated up to the moment corresponding to the maximum admissible degree of photoionization of ^{235}U isotopes. Then the radiation rapidly terminates. The ions of ^{235}U isotopes produced are pulled by an electric field to the collector plates. Unionized atoms of ^{235}U isotopes also leave (at the heat velocity) the radiation zone and are condensed on the cold walls of the heat exchanger. Then the interaction zone is again filled with a new mixture of isotopes and the whole cycle is repeated. Calculations show that the maximum performance of isotope separation is achieved at the repetition frequency of laser radiation pulses at least a few kilohertz. The choice of CVL for the pumping system is close to optimal, because high-power wide-aperture copper-vapor amplifiers at the repetition frequency of 4–6 kHz exhibit efficiency close to maximal.

(3) The duration of laser pulses should be sufficiently short, not exceeding 10^{-7} s.

The typical relaxation (de-excitation) time for the levels of uranium atoms included in the cascade ionization scheme and excited via quantum transitions induced by the absorption of dye laser radiation lies in the interval of hundreds of nanoseconds (10^{-7} s). In view of this fact, the photoionization of ^{235}U should occur at such a rate that the spontaneous relaxation from these levels would not affect their population and, finally, reduce the efficiency of the whole process of the product yield. The upper limit for the laser pulse duration can be estimated as $\tau = 150$–200 ns. The pulse duration of CVL may be as long as 60 ns, which makes it possible to use them as a high-efficiency pumping source (the actual efficiency of up to 50 %) for generating the radiation pulses of a dye laser with close durations. Action of such short-duration light pulses provides rapid and efficient photoionization of isotopes and considerably reduces dissipative losses related to the influence of interatomic spontaneous relaxation to intermediate excited states.

(4) The laser system should provide high average power of radiation at sufficiently high efficiency.

The requirements of high average power and efficiency are quite obvious and important: the first provides large weight quantities of the output product (^{235}U) and the second reduces the operation cost of the AVLIS process. High average radiation power of the laser system is needed for producing large weight quantities of the final product.

Industrial isotope separation must be profitable. Let us make numerical estimates for industrial production of uranium at various annual production values.

The ionization potential for uranium is $U_{ion} \approx 6$ eV (1 eV=1.6×10^{-19} J). One gram of final product ^{235}U comprises approximately 2.56×10^{21} atoms. If light transferred to the interaction zone is completely utilized, then the laser energy per unit weight of ^{235}U is $W = 6 \times 1.6 \times 10^{-19} \times 2.56 \times 10^{21} = 2460$ J g^{-1}.

Let us consider the cases where the total annual product is

1. $P_1 = 1000$ ton $= 10^9$ g
2. $P_2 = 200$ ton $= 0.2 \times 10^8$ g
3. $P_3 = 100$ ton $= 10^8$ g

Useful consumption of laser energy (at the efficiency of 100 %) for producing the annual amount is ($E = W \cdot P$)

1. $E_1 = W \cdot P_1 = 10^9 \times 2460$ J $= 2.46 \times 10^{12}$ J
2. $E_2 = W \cdot P_2 = 0.2 \times 10^8 \times 2460$ J $= 0.49 \times 10^{12}$ J
3. $E_3 = W \cdot P_3 = 10^8 \times 2460$ J $= 2.46 \times 10^{11}$ J

If laser operates for 10 h daily, 250 days yearly ($T = 9 \times 10^6$ s), then the approximate total average radiation power of the dye laser system is

1. $W_{\text{dye min}} = E_1/T = 2.46 \times 10^{12}/(9 \times 10^6) = 270$ kW
2. $W_{\text{dye min}} = E_2/T = 0.49 \times 10^{12}/(9 \times 10^6) = 54$ kW
3. $W_{\text{dye min}} = E_3/T = 2.46 \times 10^{11}/(9 \times 10^6) = 27$ kW

With round-the-clock operation for 300 days per year ($T = 2.6 \times 10^7$ s), we have

1. $W_{\text{dye min}} = 94.6$ kW
2. $W_{\text{dye min}} = 18.9$ kW
3. $W_{\text{dye min}} = 9.4$ kW

The calculation performed yields the minimal estimate of the average radiation power for the system of dye lasers, because the efficiencies of photoionization, separation, extraction, and the losses of laser radiation during transportation to the interaction zone are neglected.

For obtaining real estimate of the total output radiation power of the copper-vapor laser system required for pumping dye lasers, the inevitable loss of the pumping radiation (CVL system) converted into the output power of dye lasers (the actual efficiency less than 50 %) and losses on passive elements of the optical system intended for transporting light beams from CVL to dye cells should be taken into account. With all the losses taken into account (including those in evaporator–separator), the calculated total efficiency relative to the radiation energy of the CVL system is at least 5 %.

Hence, the numerical estimates for the average radiation power of the CVL system are as follows.

At 10-h daily operation for 250 days during a year:

1. $W_{CVL} = 2.7 \times 10^5$ kW/$(5 \times 10^{-2}) = 5400$ kW
2. $W_{CVL} = 0.54 \times 10^5$ kW/$(5 \times 10^{-2}) = 1080$ kW
3. $W_{CVL} = 2.7 \times 10^4$ kW/$(5 \times 10^{-2}) = 540$ kW

At 24-hours daily operation for 300 days during a year:

1. $W_{CVL} = 9.46 \times 10^4$ kW/$(5 \times 10^{-2}) = 1890$ kW
2. $W_{CVL} = 1.89 \times 10^4$ kW/$(5 \times 10^{-2}) = 378$ kW
3. $W_{CVL} = 9.46 \times 10^3$ kW/$(5 \times 10^{-2}) = 189$ kW

The values of W_{dye} and W_{CVL} are really accessible because presently CVL with a unity (that is, per active element) average output power of the order of 100–150 W and more. There are no principal difficulties for creating modular radiating units on the basis of CVL with an average output power of a few kilowatt.

2.3
Laser Complex

Modern large-scale LIS complexes utilize a united concept and are built according to the common scheme. Matter is subjected to irradiation of single-frequency narrow-band radiation of dye lasers that are pumped by copper-vapor lasers. For increasing the power of both the laser systems, the "driving generator–preamplifier–final amplifier" scheme is used (see Fig. 2.1). The laser sources, in turn, can be arranged in parallel channels. Some aspects of building laser complexes for LIS are presented in Appendix B.

On the one hand, the requirements for laser physics and scientific instrument-making industry were formed by the needs of LIS; on the other hand, the development of the element base determined the possibilities of LIS complexes. Consideration of a complex as a whole cannot be performed without considering its components. Let us dwell on this question.

2.3.1
Pumping Lasers

The principal requirements for the pumping lasers are high generation power, high efficiency, specific distribution of radiation, and a number of parameters responsible for the generation stability. Recent developments of lasers make it possible to use high-power solid-state lasers and copper-vapor lasers as the sources for pumping tunable lasers.

Gas-discharge copper-vapor lasers completely meet the requirements mentioned. With longer than 30-year history, copper-vapor lasers are still actively being developed. High-power CVL became the key instrument that provided

Fig. 2.1 Driving dye laser generator.

the creation of large-scale separating complexes. It is known that gas-discharge lasers with the active media on self-limiting transitions of metal atoms (Cu, Au, Mn, etc.) generate in a wide wavelength range covering UV, visible, and near-IR spectral ranges and have the following advantages: high average (dozens and hundreds of watt) and peak (hundreds of kilowatt) radiation power; high practical efficiency (up to 2 %); the pulse-periodic mode of operation with the tunable repetition frequency of pulses from a few to dozens of kilohertz; the short pulse duration (shorter than 60 ns); possible simultaneous multicolor generation in various spectral ranges; high optical homogeneity of active media, which provides obtaining the radiation angular divergence close to the diffraction limit in unstable cavities.

Among lasers of this class, copper-vapor lasers are distinguished by the output energy, pulse repetition frequency, spatial-temporal, spectral, and other properties of radiation.

State-of-the-art achievements and ideas in the field of copper-vapor lasers are presented in [32, 33]. The active medium of such a laser is a vapor-gas neon–copper mixture, which, being excited by a pulse-periodic gas discharge, generates the coherent radiation at two different wavelengths in the visible spectral range ($\lambda = 510$ nm and $\lambda = 578$ nm). Since the creation of CVL [79], obtaining the self-heating operation mode [31], and employment of the "generator–amplifier" system for increasing the output power of radiation [80] lasers of this kind have become an attractive instrument for pumping systems due to its high practical characteristics: high efficiency of energy conversion from electrical to the radiation form; high energy of radiation quantum; short duration of exciting radiation comparable with lifetimes of atoms in excited states; high average radiation power and high efficiency of converting CVL radiation to tunable radiation. Due to the unique parameters of the output radiation of copper-vapor lasers the latter were chosen as the basis in developing the most important component of laser equipment for the AVLIS process, namely, high-power laser systems intended for highly efficient optical excitation of the second component—the dye laser system with the tunable wavelength of output radiation.

The limiting parameters of generation obtained from a single gas-discharge CVL integrated with a gas–vacuum system are known. The average generation power is as follows:

1. 550 W in the generator mode and 615 W in the amplifier operation mode for CVL with the active zone of diameter 8 cm and length 350 cm [81, 82];
2. 325 W in the amplifier mode for CVL of diameter 8 cm and length 350 cm [84];
3. more than 250 W in the amplifier mode [84];
4. 201 W with the efficiency of 1.9 % at maximum power in the generator mode for the HyBrID laser on copper vapor with a tube of diameter 6 cm and length 200 cm [85].

Some other approaches to improvement of CVL are developed in [36, 48]: the elaboration and creation of sealed-off metal vapor lasers with enhanced specific generation energy in tubes of small diameter. The laser tube with a diameter of 3.2 cm and length of 125 cm provided the average radiation power of 55 W in the generation mode and 70 W in the amplification mode with the efficiency of up to 1.5 %. At the tube length of 153 cm, the radiation power was 80 W. In such a laser with an increased pumping efficiency due to higher voltage 30–35 kV, the average power of 100 W was obtained. Obtaining the average power of 120 W at the efficiency of $\sim 2\%$ relative to energy deposition is also reported. The laser with the channel diameter of 4.5 cm and the length of 126 cm has the average power of 74 W and at the length of 152 cm it is 90 W. Advantages of lasers [36, 48] are high efficiency of standard units, simplicity, long operation period, and a moderate cost.

Despite impressive achievements in the development of CVL-based laser complexes, their further employment in the form they were developed according to the national programs of the USA, Japan, and Great Britain becomes difficult, at least in small investigation communities. The main difficulties are the necessity of the integrated vacuum system with the refreshment of the operation mixture and the large diameter of tubes. These factors limit their employment because of the large dimensions and a high cost. One more drawback is the relatively low repetition frequency of pulses $F = 4\text{--}5$ kHz, which forces multiplication of the number of channels in LIS installations in order to provide continuous light irradiation of vapors of the separated material. At the typical velocity of vapor flux in the AVLIS method, $v = 10^5$ cm s^{-1}, the distance it covers during the pulse-to-pulse interval is $l = v/F > 20$ cm, which is unacceptable for efficient separation. The multiplication increases the cost of the LIS complex in the corresponding proportion.

The employment of large-diameter tubes results in a worse quality of the output beam due to the skin effect [86] and necessitates an expensive optics for beam transportation and conversion. The facts mentioned make it sensible to further improve CVL in view of the increasing competition of solid-state lasers. Some physical–technical problems and possible improvements of gas-discharge CVL are given in Appendix C.

The second harmonics of solid-state lasers with diode pumping seems promising for pumping dye lasers. Recent progress in physics of solid-state lasers is explained by the following factors: the employment of new active media, including those with high gain; the employment of narrow-band emitting diodes for pumping, which resulted in a noticeable enhancement of the efficiency and a lower heat load in an active medium; the employment of monolithic and semimonolithic constructions capable of combining an active medium, an optical cavity, and, in some cases, control elements in a single element, which, in turn, resulted in a considerable enhancement of the stability of output radiation; and high reliability and long operation period of emitting diodes, which is usually longer than 10^4 h. The efficiency of solid-state lasers with diode laser pumping is an order of magnitude higher than that of conventional solid-state lasers [87]. The greatest progress is observed in the development of high-power YAG:Yb and YAG:Nd lasers with diode pumping. The corresponding output power in a CW mode is greater than 5 kW [88–90]. Obtaining the average output power of 100 W for the second harmonics under the pumping of 200-W YAG:Nd laser is reported [91]. Authors in [92] reported about the average power of 451 W in a YAG:Nd laser in the IR range and 182 W in the second harmonics (green range) obtained with LBO crystal under the pumping of 355 W. The average power of 315 W in the green range was also reported [93]. In view of the typical generation efficiency $\simeq 10\,\%$ in solid-state lasers and the typical conversion efficiency of $\simeq 50\,\%$ in nonlinear crystals of the BBO type (β-BaB$_2$O$_4$) and LBO type (LiB$_3$O$_5$), one may conclude that solid-state lasers with diode pumping may become preferable sources for pumping tunable dye lasers in LIS systems. However, they are more expensive as compared to CVL, which is a limiting factor in their employment. Solid-state lasers usually generate radiation with the duration of $\tau > 50$ ns, which may be unacceptable for efficient conversion of dye laser radiation into second harmonics and for separating isotopes with a short lifetime of energy levels. Thus, solid-state lasers have not yet got the combination of the radiation parameters and cost that might provide their wide application in investigation and commercial projects.

2.3.2
Tunable Lasers

The second important component in the LIS technique is the system of dye lasers capable of efficiently converting the pumping radiation into self-radiation with the efficiency of up to $\simeq 70\,\%$. In using the cavities with diffraction gratings, dye lasers provide a wavelength-tunable radiation in a wide spectral range (dozens of nanometer). In introducing additional filters inside the cavity they may generate a radiation with an extremely narrow spectral width (the record width is $\Delta f \approx 10^2$ Hz at the central radiation frequency of $f_0 \approx 0.5 \times 10^{15}$ Hz). The factor of spectral line narrowing of 10^{-8} may be achieved with active stabilization of the cavity length, which is three orders of magnitude less than the typical Doppler broadening of the spectral line. In the AVLIS process, the radiation of dye lasers is used directly for isotope excitation and photoionization.

Tunable dye lasers are those with a liquid active medium that is a solution of organic dye excited by the pumping radiation. Presently, more than 100 dyes are known [94]. They cover the spectral range from UV to IR regions. The efficiency of converting the pumping radiation may reach 50–70 %. The principal requirements for the dye lasers employed in LIS complexes are the corresponding power and efficiency of generation, the spectral composition and the spectral width of generation, the long-term and short-term amplitude and temporal stability, and the long operation period of dye [95]. General tendencies in evolution of this type of lasers are obtaining high conversion efficiency in a possibly widest spectral range; synthesizing new dyes for particular absorption lines; and making longer the operation period of dye solutions. The maximal generation parameters for a single dye laser pumped by CVL are presently as follows [96]: in the mode of a driving generator it is 25 µJ/pulse in the range 560–800 nm at the pulse duration of 40 ns and efficiency of 5–10 %, depending on the dye; in the case of four amplifiers operating at the repetition frequency of 26 kHz and the efficiency of 50 %, the average total power is 1400 W in the range 550–650 nm.

Presently, solid-state frequency-tunable lasers, in particular, $Ti^{3+}:Al_2O_3$, lasers are interesting for use in LIS complexes. $Ti^{3+}:Al_2O_3$ crystal has the maximum absorption near $\sim$500 nm, which is convenient for pumping by radiation of CVL. Such lasers may compete with dye lasers in LIS systems due to a wide tuning range of radiation (750–1000 nm) and possible employment of second harmonics. The best parameters were obtained in [84]: the power in a CW mode is 43 W and the average power is 10 W at the wavelength $\lambda = 780$ nm and the efficiency 15 % under the pumping by CVL at the repetition frequency of 8 kHz [97].

2.4
Complexes for Laser Isotope Separation

As was already mentioned, first ideas of laser isotope separation developed in physical and industrial aspects in the framework of the national AVLIS programs as applied to uranium enrichment. In recent years, an interest arose to laser isotope separation of other elements, which either can only be produced by LIS methods, or such production seems more attractive economically as compared to other separation methods. This concerns such elements as lead [98–100], ytterbium [19], palladium [101], lithium [102], neodymium [103], thulium [104], zinc [46], and some other [18, 105].

In the framework of the AVLIS programs in the USA, Japan, and France, laser complexes were elaborated with a prospect for commercial mastering LIS technologies.

The laser system developed in Lawrence Livermore National Lab (USA) [21, 35, 37–39, 64] has record parameters. The laser complex consists of 48 CVL channels (4 blocks, 12 channels each), which pump 16 channels of dye lasers. Each CVL channel includes a driving generator, preamplifier, and three amplifiers connected in series and operates at a frequency of 4.3 kHz and power of 1500 W. The channels are triggered with the time shifts such that the total repetition frequency of the CVL system is 26 kHz. The total power of the pumping lasers is 72 kW.

A unit that includes the driving generator and several radiation power amplifiers is called the line (or section) and presents a single block of the CVL system. The driving laser (DL) of a laser line provides a high-quality beam with the power sufficient for depleting the population inversion of the amplifiers. It comprises two lasers with similar active elements; one of them (generator) performs the injection control of the other (amplifier), see Fig. 2.2. The schematic optical diagram of a driving laser is shown in Fig. 2.3. The typical parameters of such a laser are as follows: the diameter of the discharge channel is 4 cm; the active length of the operation zone is 90 cm; the electrical power consumption is 5 kW per active element; the repetition frequency of radiation pulses is 4.3 kHz; the average output generation power is 25 W; the radiation divergence corresponds to 15 diffraction angles; the half-height duration of the radiation pulse is 40 ns and the duration corresponding to 10 % level is 55 ns.

The driving unit comprises seven principal components: active elements, the unit with primary power sources, thyratron switches, low-voltage electronics for controlling the operation of thyratrons, a computer, a frame with gas–vacuum system, and the main case. All the components are arranged as replaceable modules. The unit of the driving laser is fixed and adjusted on a kinematic wall of the laser tunnel. The optical elements responsible for the injection of the control signal from the generator to the amplifier are placed in the same unit. The output beam of the driving generator is contracted by a telescopic system to the diameter

Fig. 2.2 Block of copper vapor lasers.

2.5 mm and injected into the controlled amplifier through a partially transparent meniscus and a hole in the coupling mirror. The meniscus functions in two ways: first, it is a mirror of an unstable cavity in the controlled amplifier and, second, it provides the optical matching between the wavefront of the injected input signal from the generator and the radiation field of the controlled amplifier. The injection control in the driving generator is the necessary condition for obtaining a high optical quality of the beam. In this case, with the help of lens the total 100 % radiation can be passed through the hole, whereas without injection this parameter is less than 25 %. The injection makes the half-height duration of pulses passed through the hole change from 15 to 40 ns.

The radiation of the controlled amplifier leaves the cavity, reflects from the coupling and turning mirrors, and passes to the output site of DL. Then the output beam of DL with the diameter of 4 cm is telescopically expanded at the entry to the first power amplifier to the diameter of 8 cm for filling the entire aperture and is amplified up to the average power of several hundred watt keeping high optical quality.

The amplifying cascade is made similar to DL, however, with a single active element. Three primary power supplies, 15 kW each, convert an AC voltage of $\sim$ 480 V to a DC voltage of 10 kV, thus providing the consumed electrical power of 30 kW. The constructions of DL and amplifiers comprise all necessary elements.

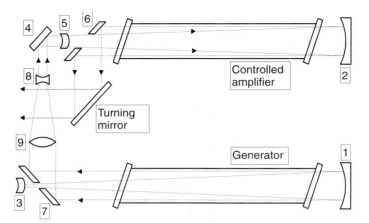

Fig. 2.3 Optical diagram of the driving copper-vapor laser:
(1, 3) mirrors of unstable cavities; (4) a small turning mirror;
(5) the partially transmitting matching meniscus; (6, 7) coupling
mirrors; (8, 9) lenses of the compressing telescope.

In order to start the operation of the cascade, it suffice to supply cooling water, buffer gas (neon), a voltage ~ 480 V, the synchronizing pulses, and attach it to a vacuum line for slowly pumping the buffer gas.

In order to increase the output radiation power of the line consisting of amplifiers with the parameters mentioned above, thorough mutual adjustment is required for the cascades, strict temporal synchronization of their operation, long-term reproducibility of the radiation parameters for each cascade, and the minimal radiation losses on the transportation and transmission optics. In 1986, a single line of laser demonstration equipment has been put into life test. At the end of 1987, the output radiation power of each of the amplifying cascades was stabilized at the level of 300 W and the total operation time of the amplifiers reached 60 000 h. The total time of the simultaneous operation of all amplifiers was 2000 h weekly.

All the amplifiers and DL have similar construction and are completely interchangeable for reducing the time needed for system repair to a minimum. If a cascade happens to break, it can be quickly replaced by a spare one. In the case of breakdown, the cascade is quickly extracted by a lifting unit; another amplifier occupies its place and warms up. In certain time lapse, the new amplifier is forced into the operation regime and supplies the system with light. The broken cascade is transported along the tunnel to a repair station for diagnostics. After the broken element is determined, it is replaced. As the heat-gas–vacuum training is finished, the cascade is ready for operation and is moved to the laser tunnel as a reserve.

The CVL radiation is transferred to the dye lasers by optical fiber guide with the transmission coefficient 80 %. Each channel of the dye laser comprises a driving generator and a four-step amplifying system with an output power of 1.4–1.6 kW at the repetition frequency of 26 kHz. The width of the laser radiation band is 50 MHz and the pulse duration is 32 ns. The total power is 24 kW. The dimensions of the AVLIS complex itself are illustrated by the area it occupies 9300 m^2, half of which refers to the laser part, and dye lasers occupy 650 m^2 [42, 106].

Up to the present time, laser AVLIS equipment in Lawrence Livermore National Lab in the USA remains the most powerful source of induced radiation in the visible range in the world.

In Japan, the laser complex provides more than 2 kW of the pumping power per CVL channel (500 W average power per single amplifier at the efficiency of 2 %; the repetition frequency is 4–6 kHz) and more than 700 W of radiation power of the tunable dye laser [24].

In the framework of the French program SILAX on uranium enrichment, a complex was created comprising a six-channel CVL system (4 driving generators and 16 amplifiers) with the total power of 2 kW and a four-channel system of dye lasers [43].

In Russia, no direct interest was shown to industrial LIS of uranium, seemingly because other methods for uranium enrichment were developed. In recent years, the ideas and approaches founded in classical works [10, 11, 14, 16–18] were developed, which was realized in the creation of moderate-size laser complexes capable of producing weight quantities of isotopes [50, 107–110]. Let us consider some of them in more details.

A system for obtaining ytterbium isotopes is described in [19, 108]. The laser complex was designed as follows. The dye lasers were pumped by copper-vapor lasers (LM-8, LM-25, LM-50) with the half-height pulse duration of 15 ns and the repetition frequency of 10 kHz. The first laser was a driving generator and the remaining lasers were used as amplifiers. Stabilized power supplies and a system of LC peakers were used, which provided high temporal and amplitude stability of the output laser pulses. The multichannel electronic unit was used for synchronizing the laser pulses with an accuracy of 1–2 ns. The average output power at the wavelength $\lambda = 510$ nm with the amplifying elements LM-25 and LM-50 was 18 W and 30 W, respectively.

The system of dye lasers included three channels made according to the scheme "driving generator–amplifier." The second amplifier was used for increasing the output power of the third channel intended for photoionization. With R6G and R110 dyes, the average output power of generation at the repetition frequency of 10 kHz was 1 W at the wavelengths $\lambda = 555.6$ nm and $\lambda = 581.1$ nm and it was 3–5 W at the wavelength $\lambda = 582.8$ nm. The half-width of first- and second-step laser lines measured by the Fizeau interferometer and a scanning interferometer was not greater than 500 MHz. The short-term instability of the wavelengths of dye lasers measured by temperature variation was approximately 400 MHz h^{-1}.

A system for obtaining palladium isotopes was considered in [109]. The experimental installation included a two-channel laser system on copper-vapor lasers and a two-channel system with dye lasers. The system with CVL included a driving generator and five amplifiers on the basis of "crystal LT-40Cu" laser tubes. It provided the total radiation power of 50 W at the wavelength of 510 nm, at the half-height laser pulse duration of 20 ns, and at the repetition frequency of 10 kHz.

The transverse pumping of dye laser chambers was used. In the first channel of laser system (R110), the narrow-band frequency-tunable laser radiation was formed near the wavelength of 552.78 nm, which, after being doubled, was used for pumping palladium isotopes. The driving generator was a commercial "LLP-504" laser with the extended-base intracavity Fabry–Perot interferometer, which reduced the spectral width of laser generation to 700 MHz at the average output power of 150 mW.

In order to reduce the spectral width of the laser radiation of the first channel and to cancel the spectrum broadening in the following frequency doubling, two narrow-band spectral filters were used, namely, two confocal interferometers with piezoelectric adjustment. One of them was placed behind the preamplifier and operated in the visible spectral range. With the base of 3 cm (the free spectral range 2.5 GHz) and at the acuity of 15, the calculated width of the instrumental function of the filter was 170 MHz. The real spectral width of the laser radiation behind the filter measured by a scanning confocal interferometer was at most 250 MHz (the half-height width averaged over 10^3–10^4 laser pulses). The decrease in the power was then compensated in the amplifier. After the frequency doubling, the spectral width became approximately twice greater.

The second interferometer was placed behind a nonlinear crystal and operated in the UV spectral range. Its base was 1.5 cm (the free spectral range 5 GHz) and the calculated instrumental function width was 64 MHz at the acuity of 78. The real spectral width of the output UV radiation was 130 MHz.

For doubling the frequency in the first channel, a nonlinear BBO crystal with the length of 7 mm was used. The average power of the narrow-band UV radiation at the output of the first channel was 5–7 mW at the conversion efficiency of up to 10 %.

The wavelength of the laser radiation was controlled in the visible range only. It was measured by a four-channel instrument with an accuracy of ± 0.00001 nm. Absolute graduation of the wavelength meter was made by the radiation of a stabilized He–Ne laser. The error in spectral wavelength measurements in the experiments was ± 0.00002 nm. In the UV spectral range (at the wavelength of ~ 276 nm) this value corresponds to the line frequency error of ± 40 MHz.

The frequency deviation of the laser radiation caused by temperature and pressure variations in ambient space was periodically compensated by manually changing the inclination of the intracavity etalon and by adjusting with piezo-drivers the cavity length of confocal interferometers to a maximal transmission.

In the driving generator of the second channel, which is similar to that of the first channel, no Fabry–Perot interferometer was used, and the spectral width of the laser radiation at the output was 15 GHz. The value of detuning of the wavelength of the UV radiation from the excited resonance transition (276.39 nm) in the second channel used for photoionization was 20–30 nm. The frequency was doubled by a BBO crystal 10 mm in length with the aperture 4×5 mm. The conversion efficiency was about 20 %, and the average power of the UV radiation at the output of the second channel was 0.5–0.7 W at a half-height spectral width of the UV radiation of 30 GHz.

The synchronization of the laser pulse passage to the zone of interaction with vapor atoms was provided by an optical delay line with an accuracy better than 1 ns. The density of the laser radiation power in the interaction zone was 10^2 W cm^{-2} for the first channel and 10^4 W cm^{-2} for the second.

Serious progress in developing the complex comprising CVL lasers and dye lasers was made at Russian Research Center Kurchatov Institute. It provided carrying out experimental operation with rare earth elements, in particular, neodymium. The complex provides obtaining an average radiation pumping power of $\sim$300 W from CVL and $\sim$100 W from tunable dye laser [110].

The laser complex for isotope separation created at Semiconductor Physics Institute, SD of RAS [50] is a combination of two independent laser channels, each generating at a separate wavelength. The corresponding functional diagram is shown in Fig. 2.4.

Driving system. The source of the frequency-tunable coherent radiation comprised CW single-frequency dye lasers (usually Rhodamine 6G) with a traditional three-mirror cavity [53] pumped by the radiation of a CW argon laser. The required total pumping power in the blue–green range did not exceed 10 W, and a typical output power of dye lasers (at the wavelengths $\lambda = 607$ nm and $\lambda = 615$ nm) was 70 mW. The cavity included three selective elements: a three-component birefringent filter, Fabry–Perot etalon, and Troitsky film. The electronic control unit provided a single-frequency mode of operation. The spectrum of the output radiation was detected by a scanning Fabry–Perot interferometer. Accurate measurements of the spectrum width (by the spectrum of beatings with two similar lasers) show that the width of the radiation line is less than 5 MHz at the measurement duration of a few seconds, and the frequency drift is a few hundred MHz per hour. The central frequency of each laser was measured by a λ-meter. The relative measurement error was 5×10^{-8}, but imperfect calibration and the temperature drift (approximately 100 MHz h^{-1}) worsened the absolute measurement error to 150–200 MHz. For the accurate adjustment to the resonance it is possible to control the frequency automatically by the signal of luminescence from the separation chamber.

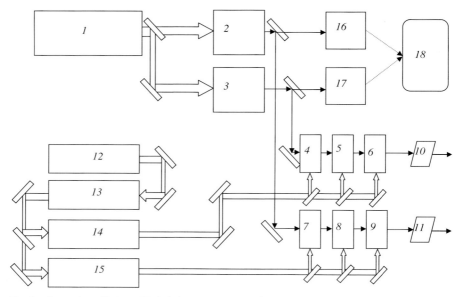

Fig. 2.4 Laser installation: (1) Ar$^+$ laser; (2, 3) CW driving generators on dye lasers; (4, 5, 7, 8) preamplifiers on dye lasers; (6, 9) output amplifiers on dye lasers; (10, 11) frequency doublers; (12) driving CVL; (13) CVL preamplifier; (14, 15) output amplifier on CVLs; (16, 17) wavelength measuring devices; (18) the control computer.

The radiation from each driving laser was amplified by a three-step pulsed system on dye lasers, which comprised cavities with transversal pumping by the CVL radiation. The path length from the driving lasers to the preamplifiers was ~10 m, which provided a small influence of the spontaneous radiation on CW lasers.

Complex of pumping lasers. The complex of the pumping lasers is based on a "driving generator–power amplifier" system with gas-discharge copper-vapor lasers. The system included a driving generator (DG), a preamplifier (PA), and two output amplifying cascades (OC). The water-cooled laser tubes were mounted on a single vertical module. For the active elements, commercial sealed-off gas-discharge metal vapor lasers were used [36, 48] (the driving generator of type LT-30Cu or GL-201, and the amplifiers of type LT-40Cu) with a discharge channel 2 cm in diameter and the lengths of 93 cm and 123 cm, respectively. The laser radiation with the divergence close to the diffraction limit was formed in DG with unstable cavity. The core radiation was selected by means of a mirror telescope and a spatial filter SF that was placed in the telescope waist. Parasitic superluminescence was eliminated by a system of additional optical cleaning of radiation. Having passed through SF, the amplified radiation is directed by the optical system to two output amplifiers.

The copper-vapor lasers are excited by the pulsed power supplies according to the scheme with the partial discharge of a storage capacitor [51]. The current was switched by vacuum tubes GMI-29A-1 (the triggered current is up to 280 A, the power is up to 5 kW, and the pumping pulse duration is $\sim$50 ns). The storage capacitor was charged from a stabilized voltage converter operating at a frequency of 30 kHz with the output voltage of up to 30 kV.

The driving generator was pumped by two different power supplies: by the thyratron source (with a TGI1-1000/25 thyratron as the switch) or by the modulator that was similar to the amplifying modulators.

The average power of DG at the output of unstable cavity with the first kind of pumping system was usually $\sim$6 W at the pulse duration of 18 ns (the saturation power at the input of PA $\sim$1 W cm^{-2}). The thyratron source of excitation introduced difficulties because of the long-term instability of thyratron triggering time ($\sim$5 ns), which resulted in the operation mistiming of the entire copper-vapor laser complex and in the reduction of the average output power in the subsequent amplifying cascades.

The average power of DG with unstable cavity pumped by the tube modulator reached $\sim$15 W at a pulse duration of 6–7 ns, which is insufficient for a complete depleting of the population inversion in PA. The use of a stretcher (for making the pulse longer) provided longer pulses of DG with the duration of up to $\sim$15 ns at an average power of $\sim$ 6–7 W at the input of PA; the output power from PA increased by 30 %. At the input PA power of $\sim$6 W, the output power was $\sim$55 W, and after two channels of OC of the CVL complex the total output radiation power in a diffraction beam was at least $\sim$150 W (75 W in each channel at the repetition frequency of 10–12 kHz, the pulse duration of $\sim$ 18–20 ns, and the average power consumed from each rectifier of $\sim$4.5 kW). By employing mirrors with various transmission and reflection coefficients it is possible to vary the power of the laser radiation at the output of each OC in certain limits. Since the duration of the generation pulse is short (10–13 ns), the source pulses for the generator and amplifier were synchronized with an accuracy of at least 0.5 ns and shifted relative to each other by a time lapse needed for light to pass from the generator to the amplifier.

A computer system controlled the functioning and the operation mode of the CVL complex.

Pulsed dye amplifiers. The pulsed amplifiers on dye solutions are two-channel systems. In turn, each of the channels includes three cascades. In dye amplifying cascades (see Figs. 2.5 and 2.6), usually ethanol solution of Phenalemine-512 was employed, which made it possible to use both CVL lines with almost equal efficiency. The distance between the cavity walls in preamplifiers was 0.5 mm and in power amplifiers it was 1 mm; the rate of dye flow through the system was 12 l min^{-1}. Two-pass transversal pumping was used. The pumping radiation was fo-

Fig. 2.5 Amplifying cascades on dye active medium (1).

Fig. 2.6 Amplifying cascades on dye active medium (2).

cused at the cavity by a cylindrical lens and then it was reflected back to the cavity by a cylindrical mirror. The concentration of the dye was chosen such that the transmission 30 % per passage was provided at the wavelength of the pumping radiation. As a result, the pumping radiation was efficiently utilized (90 %) and the beam distribution was more uniform.

For suppressing the amplified spontaneous radiation, dispersive prisms combined with a slit diaphragm were used and a lens telescope with a spatial filter having the waist of 0.2 mm. In addition, the delays between the pumping pulse and the instant of the amplified pulse passing were thoroughly adjusted. The optimal delay was $\sim$ 1–2 ns. The Glan prism was placed in front of the output amplifier, because a polarized radiation was needed in further processes. The optimal gain of the first preamplifier was 10^4, which provided the output radiation power of $\sim$50 mW under the pumping of 10 W. At higher gain, the power of superluminescence exceeded that of the amplified radiation. At the output of the second preamplifier, the power was $\sim$450 mW under the pumping of 12 W. After the output cascade, the power was 12 W under the pumping of 48 W. The output radiation had almost Gaussian distribution. Even higher power was obtained, however, at worse distribution, which reduced the power of the UV output radiation obtained by doubling the frequency in BBO crystals. Hence, the total laser system was optimized with respect to the radiation of second harmonics.

Frequency doubling system. The UV radiation was obtained by doubling the frequency of the dye laser in a nonlinear crystal. The radiation from the dye lasers was focused onto the BBO crystal 7 mm in length by a lens with a focal distance of 8.5 cm. The crystal was placed at a distance of 40 cm from the center of the cavity of the output cascade. Under these conditions, the output radiation of the dye lasers was 12 W in the wavelength range of 600–615 nm. Then, it was possible to obtain the second-harmonic radiation after doubling the frequency of the dye laser (300–308 nm; the average power was $\sim$3 W; the pulse half-height duration was 9 ns; see Fig. 2.7) with good, close to Gaussian, spatial distribution (see Fig. 2.8).

The power of the second-harmonic radiation is shown against the power of the first harmonics in Fig. 2.9, which makes it possible to estimate the conversion efficiency. The efficiency is almost linear over the whole operation range, which explains the fact that the half-height durations of the first and second harmonics are equal.

Complexes of tunable lasers combined with the frequency doubling system make it possible to obtain an extremely narrow spectrum of the output radiation. In our case, at the half-height pulse duration of $\tau = 10$ ns, the limiting spectrum width of the output dye amplifier is $\Delta\nu_{\text{lim}} \approx (2\pi\tau)^{-1} = 16$ MHz. The experimentally measured width of the radiation of the second harmonics was less than 45 MHz.

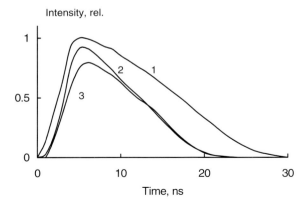

Fig. 2.7 Oscillograms of generation pulses: (1) the pumping laser; (2) dye laser; (3) second harmonics.

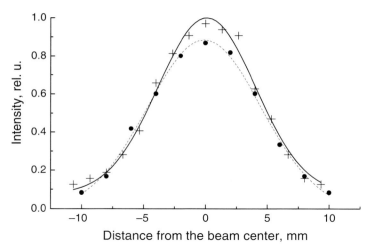

Fig. 2.8 Spatial distribution of radiation: crosses: from the output amplifying cascade on the dye laser; dots: for the second harmonics. The solid and dashed lines denote Gaussian distributions.

For long continuous operation (up to 100 h) the spectral and energy characteristics of the whole laser complex were quite stable and reproducible.

The laser complex described above is actually universal. It makes possible to carry out experiments on exciting vapors of various substances by different exciting schemes and under various conditions and to produce weight quantities of isotopes. Concrete results on experiments and producing isotopes of lead, zinc, and rubidium will be presented in Chapters 4 and 5.

Fragments of a laser separation complex are shown in Fig. 2.10.

2.4 Complexes for Laser Isotope Separation

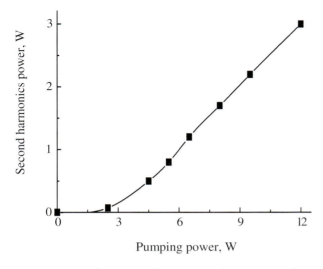

Fig. 2.9 Power of the second-harmonic radiation versus the pumping power.

Fig. 2.10 General view of separation complex.

3
Chemical Reactions of Atoms in Excited States

Atomic and molecular processes occurring at the thermal energies of impacting particles are present in all methods of isotope separation. In this chapter, we generalize the investigation results of determining the characteristic rates of such processes. The analysis is based on experimental results and the theoretical models available. In addition to being useful and important for laser isotope separation processes, there are those reducing the selectivity, efficiency of excitation, and the output of separation. We do not, however, cover all the reactions investigated up to now. We focus our attention only on those that noticeably influence the process of isotope separation.

3.1
General View of Photochemical Reactions

The experimental results, presented in monographs [70, 111], and other works devoted to the investigation of collisions of electron-excited atoms with molecules and atoms, are based on conclusions made as early as in 1952 from relatively few studies [112]. The rates of such processes depend on the energy difference ΔE between the initial and the final states of particles and have a pronounced resonance character. An almost exact coincidence corresponds to a large cross-section of impact ($\sim 10^{-13}$ cm^2). Moderate cross-sections ($\sim 10^{-16}$ cm^2) are obtained at the energy difference of a few kT (here k is the Boltzmann constant and T is the gas temperature), while the cross-sections are much lower for energy differences greater than a few decimal units electron-volt.

Chemical reactions occurring between electron-excited atoms and molecules, resulting in a stable complex, are of great importance in laser isotope separation. They may help exclude one or two steps from the process, which considerably simplifies isotope separation and makes it inexpensive.

In an impact of atom M in the ground or excited state with molecule XY ($M, M^* + XY \rightarrow M^+(XY)^-$) the following reaction channels are possible (in the order of increasing endothermic character of the reaction):

(a) $MX + Y$, k_1, k_1^*
(b) $M^+ + XY^-$, k_2, k_2^*
(c) $MX^+ + Y^-$, k_3, k_3^* (3.1)
(d) $M^+ + X + Y^-$, k_4, k_4^*
(e) $M^+ + X^- + Y$, k_5, k_5^*
(f) $M^+ + XY + e$, k_6, k_6^*

where M and M^* are atoms in the ground and electron-excited states, respectively; XY is a molecule; k_i, and k_i^* (i =1–6) are the rate constants in the ground and excited states. Ion channels (b)–(f) are usually strongly endothermic, they open at high exciting energies and fast atoms. In addition, the extraction of ions requires an electric field to be applied. This introduces certain problems, which we have already considered in Chapter 2. It is easier to carry out separation in the conditions of a gas cell with the stream of atoms and a buffer gas in the presence of a gas-reagent (XY). In the latter case, channel (a) is the main one. Nevertheless, new difficulties arise (as described later) in carrying out laser isotope separation.

If the rate of reaction (3.1) for excited atoms is higher than that for unexcited ones ($k_1^* \gg k_1$), then stable or long-living products may deposit at the walls of the reaction chamber, whereas the atoms that have not yet reacted may collect in some other place. Hence, if selectivity of excitation is provided, then the isotopes can be separated.

The general requirements for the parameters of promotional reactions (3.1) used in isotope separation are as follows:

1. The rate of reaction with unexcited atoms (the background reaction) should be of a low order for the condition $k_i N \ll 1/t$ to hold; here, N is the concentration of XY molecules in the impact zone, and t is the duration for which the atom is present in this zone.
2. All excited atoms should react chemically during their lifetime, that is, $k_i^* N > 1/\tau$ where τ is the lifetime of the excited state.
3. The reaction products MX should accumulate at the walls of the reaction chamber. Radical Y produced in the course of the reaction should react weakly with atoms M.

The choice of gas-reagent for a particular kind of atom, satisfying these conditions and providing the necessary selectivity of separation, is a hard task because complicated molecules in the reaction chamber may a number of secondary physical–chemical processes leading to lower selectivity, loss of the product and worse separation. Nevertheless, experiments show that for most of the atoms it is possible to choose the corresponding gas-reagent capable of providing the required selectivity of excitation. We will show that the cross-sections of most of the photochemical reactions are in the range 10–100 Å^2, that is, greatly exceed the gas-kinetic cross-sections. Such high rates are explained in the framework of the "harpoon model" [113]. This mechanism is based on the concept that in the approach of atom and molecule (or, in the general case, of the surface of potential energy), the covalent term crosses the ion term at point r_c (see Fig. 3.1).

At the crossing point, a nonadiabatic transition is possible: atom M casts its valence electron ("harpoon") to the nearest atom of the molecule and then pulls it by Coulomb forces. Hence, reaction (3.1) occurs through the formation of the intermediate complex $M^+(XY)^-$. At large values of r_c, the following expression holds:

$$r_c = e^2/\Delta E \tag{3.2}$$

where ΔE is determined by the difference of the ionization potential of atom M and the energy of electron affinity of the XY molecule:

$$\Delta E = I(M) - EA(XY) \tag{3.3}$$

The cross-section of the reaction in this case is given by the formula

$$\sigma \approx \pi r_c^2 \tag{3.4}$$

The affinity energy $EA(XY)$ for most of the molecules does not exceed 1 eV. Electron excitation in an atom noticeably reduces the ionization potential $I(M)$ thus sharply increasing r_c, which results in a large cross-section of the reaction. For example, in the case $\Delta E = 3$ eV, which is typical for most of the impacts we have, $\sigma \approx 60$ Å^2. The result obtained is in accordance with the experimental data.

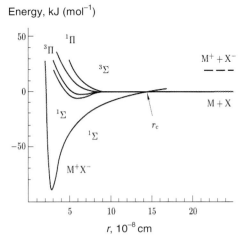

Fig. 3.1 Curves of the potential energy for alkali–haloid molecules (on the example of KBr) with the crossing ion and covalent states [113].

Other more exact models [114, 115] include a great number of covalent and ionic electron-vibrational potential surfaces in the process of electron transfer to molecules. Each of them is characterized by vibrational quantum numbers for molecules XY and XY^-. These surfaces overlap forming a grid with numerous nodes. Then, in $M + XY$ impacts, the system follows a rather complicated trajectory passing many intersections until the electron transits to an energetically favorable surface. Depending on the reaction energy, the products may be in excited electronic or vibrational states. The corresponding cross-section of the reaction is calculated by the expression [116]

$$\sigma = 2\pi r_c^2 p(1-p) \tag{3.5}$$

where p is the probability of a nonadiabatic transition at the intersection point calculated by the Landau–Zener formula. The cross-sections determined in this way are in good agreement with the experimental data. There are other models [116] that describe reactions between atoms and molecules.

3.2
Experimental Study of Photochemical Reactions Between Atoms and Molecules

In this section, we consider photochemical reactions between electron-excited atoms and molecules resulting in the formation of long-living products (see Fig. 3.2). A great number of experimental works carried out to date are devoted to investigations of chemical reactions involving excited atoms. The atoms were excited either through an interaction with resonance radiation of pulsed frequency-tunable lasers, or certain electron-excited states were formed in the photolysis of molecules by flash-lamps or high-power laser operating in the UV spectral range. In many cases, excitation in a gas-discharge plasma was used. The absolute values of the rate constant were measured by detecting the time decay of luminescence from the excited state at various pressures of the gas reagent. More informative is the time evolution study of reaction products by means of a frequency-tunable laser. The frequency of the latter should be matched with the absorption lines of the products under study. An example of the experimental setup used for measuring absolute values of the rate constants for the reaction of an $O(^1D)$ molecule with an HD molecule [117] is given in Fig. 3.3.

Under the action of the radiation of an excimer ArF-laser ($\lambda = 193$ nm), the N_2O molecule dissociated with the formation of $O(^1D)$ atoms. The luminescence was excited by vacuum UV radiation of Lyman lines H_α with $\lambda = 121.567$ nm and D_α with $\lambda = 121.343$ nm. By this luminescence, the products of the following reactions were detected:

$$O(^1D) + HD \rightarrow OD + H \tag{3.6}$$
$$O(^1D) + HD \rightarrow OH + D \tag{3.7}$$

3.2 Experimental Study of Photochemical Reactions Between Atoms and Molecules

Fig. 3.2 Photochemical reactor.

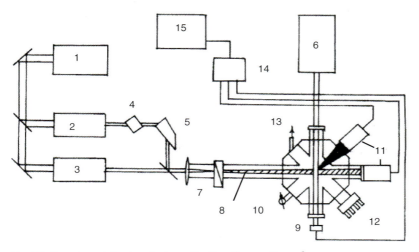

Fig. 3.3 Experimental setup for investigating the reaction of the $O(^1D)$ molecule with the HD molecule [117]: (1) XeCl-laser ($\lambda = 306$ nm); (2) dye laser A ($\lambda = 425$ nm); (3) dye laser B ($\lambda = 845$ nm); (4) nonlinear BBO crystal; (5) Pellin Broca prism; (6) ArF-laser ($\lambda = 193$ nm); (7) mixture Kr/Ar; (8) radiation ($\lambda = 121.6$ nm); (9) photodiode; (10) pressure sensor; (11) secondary emission photocell; (12) gas entry; (13) gas evacuation; (14) boxcar integrator; (15) computer.

The experiments were carried out at room temperature. For reaction (3.6), the rate constant was 1.3×10^{-10} cm^3 s^{-1}, and for reaction (3.7) it was 1.0×10^{-10} cm^3 s^{-1}.

Reactions involving atoms of alkali metals in excited electron states and hydrogen molecules were intensively studied. Collisions of atoms in highly excited electron states with H$_2$ are energy-efficient and lead to the chemical reaction

$$M^* + H_2 \rightarrow MH + H \tag{3.8}$$

Reactions of Cs atoms in 8P and 9P states [118] with hydrogen results in the origin of CsH mainly in low ($v = 0, 1$) vibrational states. Reactions of K atoms in the states 5P, 6S, 7S, and 7P yield KH molecules ($v = 0 - 3$), and a strong dependence is observed of the reaction cross-section on spin-orbit symmetry of the excited atoms [119]. Reactions of Na atoms (4P, 3P) or Rb atoms (5D, 7S) with H$_2$ were also investigated [120, 121]. The harpoon mechanism described above, which leads to a large cross-section of reaction ($\approx 10^{-15}$ cm^2) is common for all these experiments. A concurrent quenching of excited atoms occurs efficiently in addition to the chemical reaction. It is shown [120] that one half of the collisions of Na (4P) with H$_2$ leads to quenching, whereas the second half of the impacts leads to chemical reactions.

To separate isotopes by photochemical methods it is necessary for a chemical reaction to occur during the lifetime of the excited atom, that is, the inequality $k_i^* N > 1/\tau$ should hold. The rate constant k_i^* for atoms of alkali metals is equal to $\sim 10^{-10}$ cm^3 s^{-1}, and the lifetime of the excited states in the experiments discussed above does not exceed 100 ns. The maximum gas-reagent concentration is $\sim 10^{16}$ cm^3. It will be shown in Chapter 4 that at lower concentration it is very difficult to perform isotope separation. Hence, in the case considered, a chemical reaction is inefficient because less than 10 % of the excited atoms would have a chance to be involved in it. It is necessary to excite atoms to long-living states and to search for gas-reagent with the rate constant greater than 10^{-10} cm^3 s^{-1}. In the absence of metastable and long-living states, the only way to realize the conditions mentioned above for alkali metals is to excite them to long-living Rydberg states.

The rate constants were measured [122] for the chemical reactions between Rb atoms excited to the Rydberg 11P$_{3/2}$ state (with a lifetime of 0.6 µs) and molecules of diethyl ether, (C$_2$H$_5$)$_2$O, and methanol, CH$_3$OH, in photochemical reactor (see Fig. 3.3). The corresponding rate constants are $k^* = 8.4 \times 10^{-10}$ cm^3 s^{-1} and $k^* = 1.47 \times 10^{-9}$ cm^3 s^{-1}. The rate constant of the background reaction k is three orders less in magnitude than k^*. The measurements were performed by detecting the time decay of visible luminescence that arose under the pulsed excitation of Rb atom from the ground state by Rydberg radiation of wavelength $\lambda = 311$ nm at various concentrations of gas-reagent. The most probable scheme of this exothermic chemical reaction, which was also detected by the reduction of the concentration of Rb atoms in the interaction zone is the following:

$$Rb^* + G \rightarrow RbO + R \tag{3.9}$$

Hence, Rydberg states can definitely be employed for laser isotope separation by the photochemical method. In the case of Rb, the probability of reaction (3.9) is an order of magnitude greater than that of the spontaneous decay of state $11P_{3/2}$ at gas-reagent concentration $N_{gr} = 10^{16}$ cm^{-3}.

Reactions involving atoms from Group II of the periodic table of elements and certain other molecules have also been investigated. In the case of H_2 the measurements were carried out with $Mg(3s3p^1P_1)$ [123], $Mg(3s4s^1S_0)$ [124], $Hg(6^3P_1)$ [125, 126], $Zn(4s4p^1P_1)$ [127], $Zn(4s4p^3P_1)$ [128]. The products of these reactions are H and MH. The latter is produced in the excited states with large rotational quantum number and the cross-sections of the reactions are greater than the gas-kinetic cross-sections. In photochemical reactions involving $Zn(^1P_1)$ [127,130–133] or $Cd(5^3P_j)$ [129, 130, 134] atoms and saturated hydrocarbons (alkanes), the corresponding hydrides are produced in excited rotational states and radicals. The cross-sections of these reaction are greater than those of reactions with H_2.

Reactions of $Cd(5^3P_j)$ with various oxygen-containing molecules were studied in [129,130,135,136]. The cross-sections of these reactions are of the order of those for gas-kinetic reactions. The reaction of $Cd(5^3P_j)$ with H_2O yields the products CdOH and H [136]. The reaction of $Mg(3s3p^1P_1)$ with CO_2 [137] results mainly in the formation of MgO molecules. Impacts of $Zn(4p^1P_1)$ with H_2O yield ZnH in an excited rotational state [138].

In [122], the rate constant of the reaction between $Zn(4p^3P_1^0)$ and diethyl ether $(C_2H_5)_2O$ molecules was found from the rate of luminescence decay as a function of the gas-reagent concentration. The experimentally detected signal of luminescence observed after exciting the upper state by pulsed resonance radiation at the wavelength $\lambda = 307$ nm with a duration of 10 ns is shown in Fig. 3.4. The products of this reaction are ZnO molecule and a radical; the absolute value of the rate constant is $k^* = 1.61 \times 10^{-9}$ cm^3 s^{-1}. Also, reactions were investigated between $Ca(^1D_2, ^3P_j)$ [139, 140] or Ba $(6s5d^3D_j)$ [141] and RX molecules, where R=CH_3, C_2H_5, and n-C_3H_7, and X=I, Br, F. The corresponding products are MX and radical R.

The reactions of excited Al atoms with H_2 and CH_4 molecules [142] result in the formation of an AlH_2 radical and occur at low efficiency. However, it is shown [143] that reactions involving most of the atoms from Group III of the periodic table of elements, in particular, Al, and oxygen-containing molecules may also be efficient if the atoms are not excited. The latter is valid for reactions with various unexcited atoms and molecules even at room temperature [144]. Such reactions are not suitable for isotope separation.

Collisions between atoms from Group IV of the periodic table of elements, in particular $C(^1D)$ and H_2 [145] result in the formation of CH with a high rate constant that exceeds the gas-kinetic rate constants. Reactions between $Ge(4p^2 \, ^1S_0)$ and CO_2 or SF_6 molecules [146] produce GeO and GeF, respectively. The absolute values of the cross-sections of certain reactions are presented in Table 3.1.

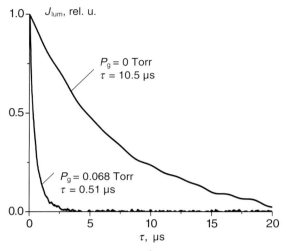

Fig. 3.4 Time decay of $Zn(4p^3P_1)$ luminescence at various pressures of $(C_2H_5)_2O$ gas-reagent; $T = 570$ K [122].

Table 3.1 Absolute values of the cross-sections for reactions between excited atoms and molecules, 10^{-16} cm^2

Molecule	Ca($^1D_2$)	Cd(^3P_j)	Rb(5d)	Na(4p)
CH$_3$I	8.0 [139]			
C$_2$H$_5$I	4.3 [139]			
C$_3$H$_7$I	2.4 [139]			
H$_2$		12.3 [134]	10 [119]	9 [120]
CH$_3$CH$_2$CH$_3$		6.6 [134]		
(CH$_3$)$_2$CHCH$_3$		1.3 [134]		

3.3
Collisional Quenching of Excited Atomic States by Molecules

The quenching of excited atomic states in impacts with molecules not accompanied by a chemical reaction plays a negative role in laser isotope separation. Such a process results in the excitation of vibrational–rotational states of the molecule. Firstly, this process leads to the broadening of resonance absorption by atoms, which in many cases reduces isotopic selectivity and the cross-section of optical excitation. Secondly, quenching removes useful excited atoms from the reaction zone, thus reducing the quantum yield of the reaction. Thirdly, the exchange of excitation energy between isotopes results in less isotopic selectivity.

The most commonly used method for determining what relative part of chemical reaction results from impacts is measuring the concentration of atoms or reaction products prior to and after the pulsed excitation of atoms. In addition, the total rate of quenching the excited state caused by the presence of molecules is measured. Experiments show that the rate of such process depends, to a great degree, on specific features of the impacting particles. For example, Table 3.2 presents the measurement results of absolute values of the quenching cross-sections and the relative yield of CdH in collisions of excited Cd (^3P_j) atoms with alkanes [134]. One can see that the relative yield of the chemical reaction producing stable compound CdH varies from 0 to 100 % depending on the type of molecule. A similar situation is observed in collisions of Ga ($5^2S_{1/2}$) atoms with various molecules (see Table 3.3) [141]. In addition to the formation of HgH, the H_2-dissociation channel is also efficient in collisions of Hg ($^3P_1$) with H_2. In this case, the quantum yield of producing HgH is 0.8 [111]. Hence, in each particular case, for the reaction to occur in the preferred chemical channel it is necessary to thoroughly choose an appropriate gas-reagent.

The absolute values of the cross-sections of some processes of quenching atomic excited states by molecules are given in Table 3.4. The average quenching cross-

Table 3.2 Quenching cross-sections and the relative yield of CdH in impacts of Cd(^3P_j) with various molecules at $T = 553$ K; ΔE is the energy deficiency of CdH formation [134]

Molecule	ΔE (eV)	Total cross-section of quenching (10^{-16} cm^2)	Relative yield
H_2	+0.0	12.3	1.0
CH_4	≈ 0	0.0041	< 0.2
CH_3CH_3	−0.22	< 0.016	> (0.3 ± 0.1)
$CH_3CH_2CH_3$	−0.35	0.0082	0.8 ± 0.3
$(CH_3)_2CHCH_3$	−0.48	0.011	1.2 ± 0.3
c-C_6H_{12}	−0.35	0.19	0.4 ± 0.1
CH_2=$CHCH_3$	−0.61	185	0.0

Table 3.3 Absolute values of the rate constant of decay of Ga ($5^2S_{1/2}$) excited state and the relative part of chemical quenching in impacts with various molecules [141]

Molecule	Rate constant for decay (10^{-10} cm^{3}s^{-1})	Relative part of chemical quenching
CH_4	3.8 ± 0.4	0.27
C_2H_6	4 ± 1	0.33
C_3H_8	5 ± 1	0.26
N_2O	4.8 ± 0.6	0.96
CO_2	5 ± 1	0.55
C_2H_4	4 ± 1	0

Table 3.4 Absolute values of the cross-sections of quenching excited states by molecules, 10^{-16} cm^2

	Ar	He	N$_2$	CH$_4$	H$_2$	O$_2$
Cd(5^3P_j)	$< 8 \times 10^{-1}$ [130]	$< 3 \times 10^{-1}$ [130]	2.4×10^{-2} [129]	4×10^{-3} [130]	12.3 [129]	
Cd(5^1P_1)	1.1 [135]	$< 3 \times 10^{-2}$ [129]	48.5 [135]		24 [129]	
Hg(^3P_j)	$< 6 \times 10^{-1}$ [130]		0.9 [70]	0.2 [130]	22 [70]	57 [70]
Zn($^1P_1$)			19 [127]	43 [127]	38 [127]	
Zn(^3P_j)			0.2 [127]	0.5 [130]	8.6 [111]	
Kr($5p^3D_3$)			10 [147]			
Rb($5P$)			22 [70]		3 [70]	84 [84]
Co(b^4F, a^2F)			24 [148]			20 [149]

section corresponds to gas-kinetic values, however, in some cases the quenching is negligible. In the next section we will present examples of an abrupt increase in the cross-sections for cases, where the frequencies of atomic electron transition and molecular vibrational–rotational transition are equal.

3.4
Resonance Transfer of Excitation in Collisions

The resonance collisions are the processes that occur according to the following scheme:

$$A^{**} + BC \rightarrow A^* + BC^* + \Delta E \tag{3.10}$$

In this case, the excitation energy is transfered from one of the colliding particles to another. Collisions of this kind differ from the quenching processes considered above by the additional condition $\Delta E \leq kT$ (the energy deficiency of the reaction should be small or comparable to the relative energy of the colliding particles). The partner BC to an electron-excited atom A may be a similar atom or some other isotope. It is important in resonance processes that the large cross-sections ($\sim 5 \times 10^{-14}$ cm^2) exceeding the corresponding gas-kinetic values might be realized [112].

Resonance processes may create considerable difficulties in the experimental realization of laser isotope separation. It was mentioned in the previous section that the efficiency and selectivity of excitation may fall and the time of interaction of the excited atoms with the gas-reagent molecules may become shorter in the case of photochemical separation. The probability of the atomic energy states to coincide with the vibrational-rotational levels of molecules is greater at greater principal quantum number because the density of adjacent energy levels increases. The energy conversion of selectively-excited Na ns ($n = 5$–11) atoms to the energy of the vibrational-rotational v_3 band of CH_4 and CD_4 molecules was experimentally studied [150]. The intensity of luminescence in various decay channels was investigated under pulsed excitation of the state under study. At small ΔE, the cross-section of the transfer increases. The resonance increase of the cross-section was observed for the changes $\Delta l = 1, 2$ in the angular momentum of sodium atoms and it was found to be small at $\Delta l = 0$. The efficiency of process (3.10) proved to be large for fundamental principal vibrations but was negligible for the compound vibration $v_1 + v_2$ of a CD_4 molecule, although the vibrational frequencies were exactly matched with the transition 5s–4p of the sodium atom. The results of the excitation cross-section measurements are given in Table 3.5 [150].

Table 3.5 Cross-sections of excitation transfer from various states of the sodium atom in impacts with CH_4 and CD_4 molecules

Na atomic state	σ (CH_4) (10^{-16} cm^2)	σ (CD_4) (10^{-16} cm^2)
5s	113±23	35±5
6s	230±22	71±10
7s	96±13	344±39
8s	76±5	105±11
9s	121±9	148±25
10s	155±20	129±34
11s	136±10	167±24
4p	72±10	71±10
5p	51±7	49±7
5d		43±14

In [151], the broadening of the two-photon resonance in Zn atoms was experimentally investigated for the transition $4s_2^1 S_0$–$6s^3 S_1$ in impacts with CO_2, CO, and NO molecules. These data provide direct information on impact cross-sections. The spectrum of two-photon absorption was detected by the luminescence that arises due to cascade decay of the upper level (see Fig. 3.5). This transition is interesting because it is included in the excitation process of Zn atoms in laser photochemical isotope separation (see Chapter 5). The cross-sections of broadening were determined from the measurements of the widths of resonances at various molecular gas pressures in the excitation zone. For CO_2, CO, and NO molecules these are 9.4 ± 2.4, 6.5 ± 1.6, and 3.9 ± 1 in units of 10^{-14} cm^2, respectively. So, large cross-sections noticeably exceeding the corresponding gas-kinetic values are explained by the accurately realized resonance of the transition $6s^3 S_1$–$5p^3 P_j^0$ in Zn atom with vibrational-rotational absorption bands of the compound vibration $v_1 + v_3$ in a CO_2 molecule and the overtones $2v$ in CO and NO molecules.

A great number of experimental and theoretical works are devoted to the study of energy transfer from excited atoms to the nearest levels in impacts with atoms in the ground state. Experimental data obtained by various authors for different

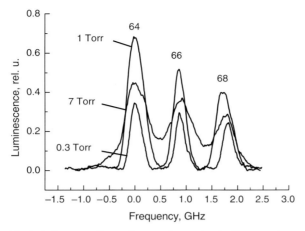

Fig. 3.5 Spectra of two-photon absorption in Zn atom in transition $4s_2^1S_0$–$6s^3S_1$ at various pressures of CO_2 molecules. Isotope numbers are designated near the corresponding peaks.

atoms [152] are shown in Fig. 3.6. One can see that the absolute values of the cross-section of energy transfer are large at small values of energy difference (energy deficiency) between the initial excited state and the state into which one of the colliding atoms transfers. One may conclude that the cross-section of transferring the excited energy is approximately 40 times greater for resonance states and falls, independent of the energy states, inversely to the energy deficiency of the reaction.

The cross-sections of excitation energy transfer to a similar atom or its isotope are greater than those presented in Fig. 3.6 because of the exact resonance. For example, the maximum cross-section of transferring resonance excitation in impacts of cesium atoms in the ground ($6^2S_{1/2}$) state with similar atoms in the resonance-excited ($6^2P_{3/2}$) states was estimated to the value $\sim 10^{-11}$ cm^2 at the velocity of particles 3×10^4 cm^{-1} corresponding to room temperature [70]. The transition under consideration has an oscillator strength of the order of unity. The cross-section of the resonance energy transfer is proportional to squire dipole momentum, hence, for weak transitions with an oscillator strength $\sim 10^{-5}$ the value would be much lower.

The processes considered in this chapter limit the concentration of atoms and molecules of gas-reagent in the interaction zone thus reducing the efficiency of laser isotope separation. For example, let us consider the conditions for separating Zn atoms (see Chapter 4). In this case, atoms are excited via the intercombinative transition $4s_2^1S_0$–$4p^3P_1^0$ with an oscillator strength of 1.4×10^{-4} and an average velocity 4×10^4 cm s^{-1} ($T = 550$ K). For accumulating the product contaminated by other isotopes by less than 1%, the concentration of Zn atoms in the separation zone should be at most 5×10^{13} cm^3.

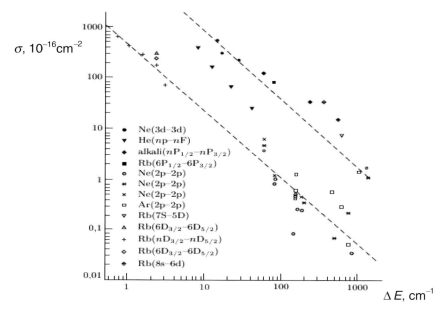

Fig. 3.6 Cross-section of the transfer of excitation energy normalized to the statistical weight of the final state in collisions of similar atoms versus the energy deficiency.

3.5
Collisional Processes with Rydberg Atoms

Highly excited (Rydberg) atomic states [153] are of great interest for laser photochemical isotope separation. Their lifetime increases with the principal quantum number ($\sim n^3$). For example, state $11P_{3/2}$ of rubidium atom has a lifetime of ~ 1.5 µs. Such a long time makes it possible to carry out efficient chemical reactions. Also, Rydberg states are considered as an intermediate step in the photoionization method of isotope separation (see Chapter 2). The large dimensions of Rydberg atoms (of the order of $a_0 n^2$) and a weak bond between the outer shell electrons and the core (≤ 0.1 eV) result in that their properties are strongly influenced by any external perturbation, in particular, by a collision with other particles.

A number of works are devoted to the broadening and shift of the exciting lines of Rydberg atoms in collisions with other atoms and molecules. Initial investigations were carried out with a high-pressure xenon lamp as the light source; the profile of the absorption line was detected by a high-dispersion spectrograph [154]. To avoid Doppler broadening, all the experiments were carried out at high pressure (~ 1 atm). Then the method based on simultaneous atomic absorption of two photons with approximately the same energy and opposite momenta was used. In this case, the profile of the absorption line is determined only by uniform broadening. The absence of Doppler broadening makes it possible to carry out experiments at rather low pressure (of the order of a few Torr). The exciting line noticeably broadens and shifts due to atom impacts with other particles. The latter fact is explained by a strong interaction of almost free electrons in a Rydberg atom with an incident particle (see, for example, [153]). The results of the investigations described in [155, 156] are given in Table 3.6.

Table 3.6 Shift Δ (MHz/Torr) and broadening γ (MHz/Torr) of two-photon absorption for transitions in Rb atom in the presence of various gases

Rb transition	(Kr) Δ	(Kr) γ	(H$_2$) Δ	(H$_2$) γ	Rb(5S) Δ	Rb(5S) γ
5S→10S	−75	240			110	320
→15S	−420	240	110	78	360	150
→20S	−485	155	135	76	127	261
→25S	−495	120	145	66	171	254
→30S	−500	115	150	65	198	252
→35S	−480	105	150	60	220	214
→9D	80	235			160	480
→10D	140	350			180	660
→15D	380	270	130	63	500	1730
→20D	445	150	140	65	970	2600
→25D	460	105	145	67	1770	2540
→30D	460	105	150	60	2020	2220

Photon echo was also used for studying line profiles. The broadening cross-sections of two-photon resonances 3S–nS and 3S–nD for collisions of a sodium atom with Xe atoms [157] obtained by this method are shown in Fig. 3.7.

The broadening process considered, generally speaking, does not lead to decay of states if it is not accompanied by other channels of relaxation or ionization from the excited states. Nevertheless, one can see from Table 3.6 and Fig. 3.7 that the absolute values of the broadening cross-sections are large, which lead to a reduction in excitation selectivity and efficiency. These values grow rapidly with the principal quantum number of the excited state and upon collisions between atoms of the same kind. In choosing the conditions for laser isotope separation, these factors should be taken into account.

Impacts of atoms in Rydberg states with thermal-energy molecules may lead to the following additional processes [153]:

$$\begin{aligned}
&(1)\ A(nl) + BC \to A(nL') + BC + \Delta E \\
&(2)\ A(nl) + BC \to A(n'L') + BC + \Delta E \\
&(3)\ A(nl) + BC \to A^+ + BC + e + \Delta E \\
&(4)\ A(nl) + BC \to (ABC) + e + \Delta E \\
&(5)\ A(nl) + BC \to A^+ + BC^- + \Delta E \\
&(6)\ A(nl) + BC \to A^+ + B^- + C + \Delta E
\end{aligned} \quad (3.11)$$

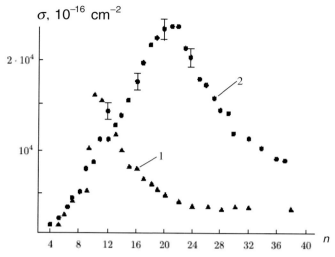

Fig. 3.7 Cross-section of broadening two-photon resonances; (1) 3S–nS and (2) 3S–nD of sodium atom in impacts with Xe atoms versus the principal quantum number.

The first two processes result in mixing the angular momenta and changing the principal quantum number of a Rydberg atom. The impacts considered are highly efficient [153], which is demonstrated in Fig. 3.8, in which the investigation results are shown for impacts between Na ($n\ ^2$D) atoms and N_2 or CO molecules [158]. If the excitation energy is not transferred to other isotopes in such impacts, then they have no negative influence on laser isotope separation because the absolute values of the rate constants of photochemical reactions and the photoionization cross-sections weakly depend on l and n.

The ionization channels 3–6 involving Rydberg atoms are rather efficient. For example, the rate constant of chemiionization, namely, processes 3 and 4 in impacts of Rb ($n\ ^2$P) atoms with Rb (5 ^{2}S) atoms [159] (see Table 3.7) strongly depends on n and is maximal at $n = 11$, which corresponds to the ionization cross-section $\approx 10^{-14}$ cm^2.

Even more efficient is process 5, which occurs due to impacts of Rydberg atoms with molecules possessing a large cross-section of slow-electron attachment. The absolute value of the ionization rate constant in the case of SF_6 molecule [153] is approximately 3×10^{-7} cm^{3}s^{-1}. Usually these values are independent of the kind of atom; they are rather related to the electron bond energy of the latter. Such fast ionization process itself may become useful for isotope-selective ionization with further ion extraction on a collector in an electric field. But in photochemical separation they play a negative role because a part of the laser excitation energy is lost in a useless channel.

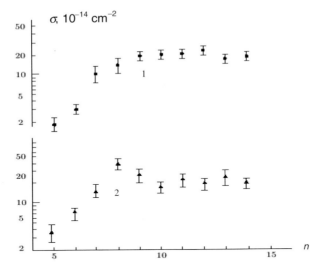

Fig. 3.8 Cross-section of mixing the rotational momenta of Na ($n\,^2D$) atoms in impacts with (1) N_2 and (2) CO molecules versus principal quantum number.

Table 3.7 Rate constants of chemionization in the impacts of Rb($n\,^2P$) with Rb($5\,^2S$); $T = 520$ K

Number n of Rb ($n\,^2P$) level	k (10^{-10} cm^{3}s^{-1})
7	$\leq 6 \times 10^{-2}$
8	3
9	7
10	9
11	14
12	12
13	11
14	9

The most serious process that may noticeably hinder selectivity and absolute yield of the separation with Rydberg atoms is the exchange of excitation energy between isotopes. This was reported [160] while measuring the rate constant of the excitation transfer between ^{3}He and ^{4}He in the following reaction:

$$^3\text{He}(n=9) + {}^4\text{He}(n=1) \rightarrow {}^4\text{He}(n=9) + {}^3\text{He}(n=1) + \Delta E \tag{3.12}$$

where $\Delta E \approx 1.5$ cm^{-1}. In the experiments, the metastable state was selectively excited by pulsed laser at room temperature. The signal of fluorescence was detected with the temporal and spectral resolution sufficient for obtaining the rate constant of process (3.12), which was found to be $(5.7 \pm 1.0) \times 10^{-10}$ cm^3 s^{-1}.

This value is very close to the rate of the charge exchange (from ^{3}He to ^{4}He) process 5.3×10^{-10} cm^3 s^{-1}. The authors concluded that the general mechanism of processes of type (3.12) is the interaction between the core of a highly excited atom and a neutral atom without participation of a Rydberg electron. Hence, the transfer of excitation between isotopes occurs as the reaction of resonance exchange of positive ion on proper atom. It is known [161] that the cross-section of such reactions are much greater than the corresponding gas-kinetic values. For example, for rubidium with the ion energy of 0.1 eV the charge exchange cross-section is 4.5×10^{-14} cm^2.

The position of energy levels in a Rydberg atom is very sensitive to external fields. The polarizability under the action of an electric field is proportional to n^7. For example, at the electric field intensity $E = 7$ kV cm^{-1}, the shift of the ground state (3s) of sodium atom is 1 MHz. For the state (30s), the shift of such magnitude is obtained at $E = 1$ V/cm. The polarizability induced by a magnetic field is far less. Hence, in laser isotope separation with Rydberg atoms an influence of external fields should be avoided.

3.6
Isotope Exchange Reactions

An isotope exchange reaction is a reaction in which isotopes of one molecule (or atom) swap places while interacting with another particle according to the scheme

$$A^{(1)}B + A^{(2)}C \leftrightarrow A^{(2)}B + A^{(1)}C \tag{3.13}$$

Molecules may be in the gaseous, liquid, or solid state and form clusters or are adsorbed on the surface of a separation chamber. Processes of isotope exchange (3.13) play a negative role in laser isotope separation because they reduce selectivity. Indeed, for the contribution to selectivity loss to be less than 1% at the pressure of reacting particles 1 Torr and the duration of the interaction 1 s, the rate constant of reaction (3.13) should not be greater than $k \approx 10^{-19}$ cm^3 s^{-1}.

In many cases, the rate constants of an isotope exchange reaction depend on temperature and are described by formula [162]

$$k = A \exp(-E_a/kT) \tag{3.14}$$

where A is a constant and depends on the kind of impacting particles; E_a is the activation energy of the process, which is usually a few tenths electron-volt. Experimental results show [162] that the absolute values of rate constants for most of the reactions at room temperature are of the order of $\sim 10^{-20}$ cm^3 s^{-1}. For example, in the experimental study of the exchange reaction between nanoparticles and a gas that proceeds according to the scheme

$$^{235}UF_5(\text{particle}) + {}^{238}UF_6(\text{gas}) \leftrightarrow {}^{238}UF_5(\text{particle}) + {}^{235}UF_6(\text{gas}) \qquad (3.15)$$

the rate constant of the reaction is $k = 5.5 \times 10^{-20}$ cm^3 s^{-1} [163].

The rate of isotope exchange reactions increases with temperature and in the presence of a surface catalyst [164], for which purpose the walls of the interaction chamber may be used. The reaction is also fast if an efficient chemical reaction producing intermediate complexes participates in the process. The latter effect was demonstrated in [165] while studying the exchange of ^{13}N and ^{14}N isotopes in the interaction of NO_2 and N_2O_5 molecules. The exchange through the reversible chemical decay of N_2O_5 to NO_2 and NO_3 has been settled with high probability. The reverse reaction yields trace amounts of nitric pentoxide. The rate constant of this reaction is of the order of 10^{-18} cm^3 s^{-1}.

In studying the exchange reaction between OH$^-$ and D$_2$, and between OD$^-$ and H$_2$, a considerable influence of the rotational excitation energy of impacting molecules was observed [166]. The rate of exchange reactions increases noticeably in the presence of a gas discharge involving the reacting particles. The latter was experimentally demonstrated with examples of reacting molecules (H$_2$O, D$_2$) and (D$_2$O, H$_2$) [167]. In isotope exchange reaction [168], in a flux with electrical discharge, the following absolute values for the rate constants at the temperature of 298 K have been obtained: for reaction ^{18}O and NO it is $k = 3.7 \times 10^{-11}$ cm^3 s^{-1}; for reaction ^{18}O and O$_2$ it is $k = 2.9 \times 10^{-12}$. Such high values are explained by additional processes that occur in a gas discharge plasma, for example, electron and vibrational excitation, ionization and dissociation, etc. An increase in the reaction rate constant was also experimentally observed in the presence of microwave radiation [169].

There is no information about carrying out experiments involving isotope exchange reactions in the presence of selectively excited electrons or vibrational states of reacting particles. Based on the experimental results presented we may conclude that the rate constants in such processes may be comparable to those discussed in previous sections of the chapter.

We may now conclude that isotope exchange reactions (3.13) may play a significant negative role in realizing laser isotope separation with the photochemical method, because they reduce the selectivity of the process. This fact explains the results of unsuccessful experiments. To get better selectivity it is necessary to investigate the gas-reagent with which the reactions considered become insignificant.

3.7
Radical Reactions in Collisions

A photochemical reaction between excited atoms and molecules results in free radicals. In impacts with molecules and atoms they induce secondary reactions, which may result in reduced selectivity and efficiency of laser isotope separation. The high chemical activity of radicals is explained by the incompletely filled electron shells in the corresponding atomic ensembles, due to which their properties become similar to those of atoms possessing the same number of outer shell electrons. An analogy exists between the chemical properties of carbon, nitrogen, oxygen, and fluoride hydrides and those of atoms with the same number of outer shell electrons [170]. For example, the CH radical is chemically similar to nitrogen atom, while CH_2 and NH radicals are analogs of the oxygen atom. Due to chemical unsaturation of radicals, the activation energy of the processes involving such radicals is of the order of the activation energy of atomic reactions. This is why such processes occur at approximately the same rate as atomic processes. Also, radicals resulting from a photochemical reaction may have electron and vibrationally-rotational excitation, in which case their chemical activity sharply increases. Secondary radicals may also arise, however, with lower chemical activity [170].

Reactions between radicals and other particles have been studied for a long time by various methods [171]. Typical values of the rate constant are in the range from 10^{-14} to 10^{-10} cm^3 s^{-1}; the activation energy is from a few tenths to 1 eV [171–173]. CH is one of the most chemically active radicals; in impacts with CH_4 and H_2S molecules its rate constants are 7.6×10^{-11} cm^3 s^{-1} and 2.8×10^{-10} cm^3 s^{-1}, respectively [174]. Much greater rates are observed in impacts of two radicals. This was experimentally demonstrated taking the reaction between $OH(X^2\Pi)$ and $NH(a^1\Delta)$ [175] with the rate constant 1.2×10^{-9} cm^3 s^{-1}, as an example.

Hence, reactions with the formation of free radicals may deteriorate laser isotope separation by the photochemical method. The strong influence of such processes may be avoided by carefully choosing reagents, with which free radicals produced in the course of chemical reactions possess low activity in impacts with other particles in the separation zone. It is also possible to add specific acceptors that would bind free radicals. The latter method was experimentally realized in the photochemical separation of mercury isotopes [176] and in multiphoton dissociation [177].

4
Isotope Separation by Single-Photon Isotope-Selective Excitation of Atom

4.1
Description of the Method

In 1932, well before the discovery of lasers, the idea of separating isotopes by the photochemical method was proposed [2]. The idea was to irradiate a mixture of mercury vapors and oxygen using a mercury lamp, whose radiation was adjusted by means of a magnetic field to predominantly excite atoms of a single isotope of mercury. In the late 1950s, a sufficiently large number of molecules were found, which chemically interact with mercury atoms excited to resonance states. The list of reagents includes such gases as N_2O, H_2O, O_2, HCl, CH_3Cl, CCl_4, CH_2Cl_2, CH_3Cl_3, and so forth [178]. Butadiene, propylene, and benzol were used as acceptors of free radicals. With the installation of "Photon-M" developed at the Russian Research Center Kurchatov Institute [179] mercury isotopes were obtained in quantities exceeding Russian needs.

Nevertheless, isotope separation with lamps has serious drawbacks. It can be only applied to a limited number of chemical elements. Mercury is the ideal element in this sense. Mercury lamps at the resonance line ($\lambda = 253.7$ nm) exhibit sufficiently high radiation efficiency. The relatively high lifetime (more than 100 ns) of resonance levels, as compared to other elements, promises a high efficiency of chemical reaction. Also, due to the high cost of mercury isotopes and low requirements, the efficiency and economic feasibility of isotope separation are not of primary importance.

The use of frequency-tunable laser radiation with a narrow emission line and a high specific power makes it possible to selectively excite isotopes in lengthy chambers with an efficiency a few orders in magnitude higher than that with lamp pumping. Laser radiation may efficiently excite long-living atomic levels (including Rydberg states) with the lifetime of up to 100 µs and longer. In such conditions, the choice of a proper chemical reaction is easier.

Laser Isotope Separation in Atomic Vapor. P. A. Bokhan, V. V. Buchanov, N. V. Fateev,
M. M. Kalugin, M. A. Kazaryan, A. M. Prokhorov, D. E. Zakrevskiĭ
Copyright © 2006 WILEY-VCH Verlag GmbH & Co. KGaA, Weinheim
ISBN: 3-527-40621-2

Laser isotope separation by conventional methods [18] for many elements, in particular for zinc atoms, is rather difficult. Firstly, the isotopic splitting is smoothed by the Doppler profile of the absorption line. In such cases, the photoexcitation selectivity is usually enhanced by known methods of making absorption profiles narrower: the method of the atomic beam and two-photon excitation in a standing light wave. Unfortunately, for some atoms these methods are inefficient. On the one hand, atoms are poorly accommodated even on cold surfaces, and the well collimated atomic beam can hardly be obtained [180]. On the other hand, two-photon excitation requires two powerful tunable sources of radiation in the UV spectral range [46], which complicates and raises the price of the experimental installation for laser isotope separation. Secondly, even at high selectivity and efficient excitation of low working levels, it is not always possible to find acceptable ways of ionizing or pumping the desired isotope to higher excitation levels due to the absence of nearby autoionization states.

Laser radiation employed in the photochemical method is efficient for some elements even if the corresponding isotope structure is masked by the Doppler profile. In this case, the following feature of the Doppler profile is used: the selectivity of excitation increases at greater simultaneous frequency detuning from the lines of all isotopes. The Doppler spectra of two isotopes shifted by a half width of the Doppler profile ($\Delta \nu_D$) are shown in Fig. 4.1. Such isotope shifts are frequent in practice. One can see that the ratio of excitation probabilities for exciting different isotopes by a narrow-band line increases with the detuning. For example, at the detuning equal to a half Doppler width, the ratio is close to 10. It is interesting that the selectivity in similar cases for lines with the Lorentz profile falls asymptotically.

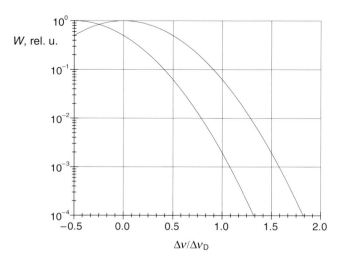

Fig. 4.1 Shifted Doppler profiles of the excitation probability for atoms.

The width of the laser radiation is usually less than the nonuniform width of the Doppler profile of transition. A single pulse of narrow-spectrum radiation excites a small part of atoms, especially if a tail of the Doppler profile is excited. Nevertheless, a great number of pulses may completely excite and "burn out" the isotope with the line nearest to the radiation frequency. It is possible to provide conditions in which the concentration of the other isotopes would not fall noticeably. The repeated action of radiation on atoms is realized at relatively low longitudinal or transversal circulation of the mixture comprising the initial isotope, the gas-reagent, and the buffer inert gas. During a single passage of the gas mixture through the active zone, the chemical reaction may simultaneously "burn out" several proper isotopes keeping the rest of the isotopes undisturbed.

4.2 Mathematical Model of the Method

In what follows we will assume a noncoherent character of interaction between radiation and atoms. This implies that at least one of the following three requirements is fulfilled:

1. The width of the radiation line is much greater than the inverse pulse duration.
2. The degree of excitation of atoms that fit into the radiation line profile is far less than unity.
3. The specific time of transverse relaxation (depolarization of medium) is far less than the pulse duration.

We will consider excitation at the pulse repetition frequency in the range 1–20 kHz and a pulse duration from 10 to 100 ns. Such a mode is specific for pumping tunable dye lasers by a copper-vapor laser. Pumping by the second harmonics of an Nd:YAG-laser is also possible.

For an efficient chemical reaction it is required that the atoms are in the excited state for a long time sufficient for a collision with a molecule to occur followed by a chemical reaction. At the concentration of reagent molecules of the order of 10^{16} cm^{-3}, the characteristic lifetimes of the levels are at least in the microsecond range. In this case, spontaneous decay during the radiation pulse may be neglected.

For the radiation pulse period, initial equations reduce to the known equations for radiation transfer and to kinetic equations. The fundamental variables are the spectral photon-flux density I_ν (the number of photons of frequency ν passing an area of 1 cm^2 in 1 s) and the spectral density of atoms $n_\nu = n(v) \cdot c/\nu_0$, where $n(v)dv$ is the concentration of atoms in the velocity interval dv and ν_0 is the transition frequency. For simplicity, index ν is omitted in the equations presented below.

The original equations in a slightly modified form are as follows:

$$\frac{\partial I}{\partial t} + c\frac{\partial I}{\partial z} = -cI\sum_i k_i m_i, \tag{4.1}$$

$$\frac{\partial m_i}{\partial t} = \beta_i k_i I m_i \tag{4.2}$$

$$n_i = \frac{m_i + (\beta_i - 1)n_{0i}}{\beta_i} \tag{4.3}$$

$$\beta_i = 1 + \frac{g_{1i}}{g_{2i}} \tag{4.4}$$

where m_i is the parameter related to the varying population n_i of the fundamental state and to the total concentration n_{0i} of levels of the ith isotope in accordance with formulae (4.3) and (4.4) (g_{1i} and g_{2i} are the statistical weights of the lower and upper levels, respectively).

The coefficient k_i is determined by the formula

$$k_i = \frac{1}{8\pi}\frac{\lambda^2}{\tau_{ci}} \tag{4.5}$$

where τ_{ci} is the radiative lifetime of the level of the corresponding transition and λ is the wavelength.

In the case of collision or power broadening of the absorption line, the profile $I(\nu)$ is a convolution of the radiation line profile with the corresponding Lorentz profile.

By integrating differential equations (4.1) and (4.2) we arrive at the following set of equations:

$$\frac{dW(z)}{dz} = \sum_i \frac{1}{\beta_i}(1 - \exp(1 - k_i\beta_i W))n_{0i} \tag{4.6}$$

$$n_i(z) = n_{0i}(z)\exp(-\beta_i k_i W(z)) \tag{4.7}$$

where $W(z)$ is the number of photons of a single pulse per unit frequency interval passed through cross-section 1 cm^2 in area at coordinate z. It is important that after the passage of the pulse, the state of the medium depends on the radiation energy rather than the pulse shape. Equations (4.6) and (4.7) will be used in the calculations given below.

It is worth noting that the parameters β and k may only differ slightly for the components of hyperfine splitting. In many cases, this difference is insignificant for the isotope separation process. For example, the main isotopes of zinc have no hyperfine splitting structure and the distance between the components of hyperfine splitting of boron isotopes is an order of magnitude less than the isotopic shift. Selective excitation mainly depends on the detuning of the radiation frequency from the transition frequency and slightly depends on the statistical weight ratio and the difference in the absorption cross-sections. Under the assumption that all the parameters mentioned above are the same for all isotopes, it is possible to obtain an exact analytical solution of Equations (4.6) and (4.7). By using the solution for m given in [181] and formula (4.3), we find the expression for the variation in concentration of atoms in the ground state for a particular isotope (index i is omitted for the sake of simplicity):

$$\Delta n = -\frac{1}{\beta} n_0 f(z) \qquad (4.8)$$

where

$$f(z) = \frac{\exp(\beta k W_0) - 1}{\exp(k p(z)) + \exp(\beta k W_0) - 1} \qquad (4.9)$$

$$p(z) = \sum_i \int_0^z n_{0i}(z) dz \qquad (4.10)$$

In formula (4.9), W_0 is the number of photons in the pulse per unit frequency interval and unit cross-section area. The sum is taken over all isotopes. Hence, there is no need to solve transport equations and kinetic equations for the time lapse of pulse action. The period between pulses is described by a system of ordinary kinetic equations. Chemical reactions, spontaneous decay of levels, and exchange reactions should be taken into account in those equations. The necessity of affecting the atom in a wing of the Doppler profile, which follows from the requirement of high selectivity results in only a small part of each isotope being excited.

$$\Delta N = -\frac{1}{\beta} \alpha f_0(z) N_0 \qquad (4.11)$$

Hence, thousands of pulses should be included in the calculation in determining variations of population level at different points and radiation frequencies. The calculation of numerous integral characteristics in performing optimization consumes much computational resources. Under certain conditions, however, the situation may be simplified. Let us move on to the concentrations integral over the atomic ground state frequencies. The variation in the ground state population in the cross-section with coordinate z integrated over frequencies can be conveniently presented in the form (4.11), where f_0 is calculated by using formula (4.9) at the central frequency of the radiation profile. The factor α introduced into this formula is responsible for the degree of isotope excitation.

$$\alpha = \frac{\int_0^\infty f_\nu n_0(\nu) d\nu}{f_0 N_0} \qquad (4.12)$$

In the case of saturated absorption we have $\alpha = 1$. If the degree of isotope excitation is small, then α is close to zero.

Assume also that the population level is affected only by the chemical reaction involving atoms in the excited state with the characteristic time τ_{char} and by spontaneous emission with a lifetime τ_{spont}, both these times being far shorter than the interval τ_f between the pulses, that is, the relation $\tau_{char}, \tau_{spont} \ll \tau_f$ is valid. In this case, by the next radiation pulse, the atomic concentration would change by somewhat less than it follows from formula (4.11), because a part of the excited atoms transfer to the ground state. In this case we have

$$\Delta N = -\frac{1}{\beta} \alpha N_0 f_0(z) \frac{\tau_{spont}}{\tau_{char} + \tau_{spont}} \tag{4.13}$$

It was mentioned that only a small fraction of atoms is excited during the time lapse of pulse action, hence, the evolution of N in a time scale much longer than the interval τ_f between pulses can be presented in the differential form:

$$\frac{\partial N}{\partial t} + V \frac{\partial N}{\partial z} = -\frac{1}{\beta \tau_f} \alpha N_0 f_0(z) \frac{\tau_{spont}}{\tau_{char} + \tau_{spont}} \tag{4.14}$$

where V is the velocity of the gas flux. Equation (4.14) allows for the transportation of atoms due to longitudinal gas circulation. A uniform plane transversal flux is described by the same equation, however, with the longitudinal coordinate z on the left-hand side substituted by a transversal coordinate. In this case, a two-dimensional distribution of atomic concentration is obtained. The transient period for the stationary concentration is of the order of the time of gas passage through the active medium, and is far shorter than the technological time of the process; hence, the time derivative in (4.14) may be neglected.

Equation (4.14) is valid for the atomic concentration of each isotope. The separation selectivity is provided by a difference in the values of α for different isotopes. If we transfer in (4.14) the dimensionless coordinate $z' = z/L$ and the reduced atomic concentration $\eta = N(z)/N(0)$ (here, L is the length of the active zone and $N(0)$ is the atomic concentration of a particular isotope at the entrance of the active zone), then it is obvious that the solution depends on the parameters $LN(0)$ and V/L (in the case of a transversal flux it depends on V only). Hence, the concentration of atoms in such regimes may be varied by changing L and, for a longitudinal flux, by changing V as well.

The choice of radiation energy for efficient isotope separation should be based on the following requirements. The radiation energy should not exceed the saturation threshold along the whole of the active zone. If it is not the case, medium becomes transparent and the radiation does not excite the selected atomic isotope. Thus, the radiation energy is wasted and the process selectivity becomes worse because undesirable atomic isotopes, not yet saturated, are excited. Less is the radiation energy, higher is the excitation selectivity. On the other hand, the requirement of high productivity forces one to increase the radiation energy at the sacrifice of selectivity. In each particular case a compromise should be chosen between the two conflicting requirements: good selectivity and high productivity.

Variations in selectivity at low or high radiation energies can be illustrated by taking an example of a weakly absorbing medium (thin optical layer). In the low-energy limit, the parameter $f_0(z)$ in (4.9) is constant and equal to $\beta k W_0$. The parameter α in (4.12) relates to the degree of overlapping of the radiation profile and the absorption profile of a particular isotope. It is proportional to the weak-signal absorption coefficient. In (4.14), α is independent of the coordinate. Then, in the stationary case the following relation can be obtained between the reduced concentrations of the ith and jth isotopes at an arbitrary point in the active zone:

$$\eta_i = \eta_j^{\alpha_i/\alpha_j} \tag{4.15}$$

Assume that the ratio of the number of atoms excited by a single pulse for two isotopes is $\alpha_i/\alpha_j = 10$ (the same as the ratio of the absorption coefficients for weak signal). Assume that the atomic concentration of the jth isotope at the end of the flux becomes two times lower; then the atomic concentration of the ith isotope would reduce by a factor of 2^{10}. Hence, the ith isotope is completely "burned out", whereas the losses of the jth isotope are acceptable. If the excitation selectivity is 10^3, which is an ordinary value in the AVLIS method and the loss of isotope is 20 %, then, according to formula (4.15), the total selectivity is 10^8. If the overlapping of the Doppler profiles with the radiation profile is weak, then undesirable isotopes are not actually excited. These conditions can easily be realized in experiments by choosing sufficiently low atomic concentrations and radiation energy. The ratio of the absorption coefficients varies over a wide range with change in the detuning of radiation frequency.

In the high-energy limit, the term $\exp(\beta k W_0)$ prevails in both the numerator and denominator of Equation (4.9); we can then assume that $f_0(z) = 1$. From this follows, according to formula (4.12), that $\alpha_i = 1$ and $\alpha_j = 1$. Hence, according to (4.15), the reduced concentrations of isotopes are equal and there is no selective excitation.

Hence, at prescribed geometrical dimensions of the active zone, the selectivity of isotope separation may be arbitrarily large, however, at low productivity. At a prescribed atomic concentration and selectivity, the productivity of isotope separation is limited above independently of the flux velocity.

In (4.14), no resonance transfer of the excitation energy or isotopic chemical exchange are taken into account (see Chapter 3). These reactions limit the admissible atomic concentration above. They definitely deteriorate the excitation selectivity. The influence of such reactions can be avoided by reducing the atomic concentration. For keeping sufficiently high absorption, it is necessary to increase the length of the active zone. The cross-section of excitation in the dipole–dipole approximation is inversely proportional to the lifetime of the upper level [8]. Since we are interested in the long-living states with the lifetime of the order of a few microseconds and longer, we may hope that the cross-section would be far less than that in typical conditions of the transfer of excitation from a resonance level, in which case it is of the order of 10^{-13} cm^{-2}. It was shown above that the atomic concentration is limited by the value $(1-5) \times 10^{13}$ cm^{-3} (see Chapter 1). A particular value of atomic concentration depends on the selectivity required.

4.3
Calculation Results on Isotope-Selective Excitation of Zinc Atoms

In this section we calculate the parameters of laser isotope separation taking the example of zinc. Its natural mixture consists of even isotopes ^{64}Zn (48.6%), ^{66}Zn (27.9%), ^{68}Zn (18.8%), ^{70}Zn (0.6%), and the odd isotope ^{67}Zn (4.1%). The corresponding isotopic shifts for the main isotopes are shown in Fig. 4.2. The Doppler width of absorption lines is noticeably greater than the characteristic isotopic shift. Hence, the isotopic structure is masked by the Doppler profile (see Section 4.6).

Calculations were performed for the transversal and longitudinal gas fluxes (see Fig. 4.3) and gave similar results. It was assumed that the profile of the radiation line is rectangular in shape. The calculation results show that the "burning out" process is highly efficient for isotopes with lines nearest to the radiation line. For example, at the radiation power of 1 W for reducing the atomic concentration of ^{68}Zn by a factor of 10, 2.5 quanta are required per single output atom. The concentration of ^{66}Zn simultaneously reduces by 30% and that of ^{64}Zn by 5%.

The method under consideration is especially efficient for obtaining isotopes whose line is extreme in the spectral composition. In a numerical simulation at the radiation power of 2 W, ^{64}Zn isotope was obtained with the fouling factor of 1% with respect to other isotopes. The energy cost of obtaining this isotope was nine quanta per atom. The total cost of separation in the complex described in Chapter 2 is \$400 per gram. Nevertheless, if the isotope radiation line resides in the middle of the spectrum, then the energy cost rises because several separation cycles would be necessary. In this case, for obtaining ^{66}Zn with the fouling factor of 1%, the calculations yield 50 quanta per atom.

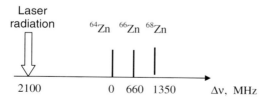

Fig. 4.2 Isotopic shifts of the main isotopes of zinc and the shift of the center of radiation line relative to the resonance frequency of isotope ^{64}Zn.

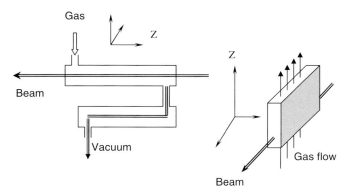

Fig. 4.3 Longitudinal and transversal circulation of a gas mixture.

4.3.1
Transversal Gas Circulation

The positions of atoms along the gas flux may be conveniently characterized by the number of radiation pulses they were subjected to. A typical dependence of the concentration of isotope atoms versus the z-coordinate along the chamber axis (to be more precise, versus the parameter $N_0 z$, where N_0 is the total initial atomic concentration) for various numbers of pulses is shown in Fig. 4.4 (the flux direction is normal to the z-axis).

One can see how the coordinate profile of the concentration of ^{64}Zn and ^{66}Zn isotopes deforms after passing every 50 pulses. A uniform distribution is assumed at the flux input. Then, as gas moves, the gradual blooming of the medium is observed due to the "burning out" of ^{64}Zn atoms. The preferable excitation of this isotope is explained by the fact that the radiation frequency was shifted from the isotope line by -2.1 GHz and from the line of ^{66}Zn isotope the shift was -2.76 GHz at the radiation spectrum width of 500 MHz. The principal characteristics of the example considered are presented in Table 4.1.

At a width of irradiation zone 2 cm, the data presented correspond to the flux velocity of 0.5 m s^{-1}.

A drawback of transversal gas circulation is a strongly nonuniform distribution of the residual "burnt" isotope along the chamber axis. The method scheme with gas circulation along the chamber axis has no such drawback. It can be avoided by using counter-propagating radiation beams.

There are two possible variants of using counter-propagating beams. In the first case, it is sufficient to place a reflecting mirror at the output of a chamber. In the second case, an additional laser may be used. The second variant seems preferable, because in the case of medium blooming the radiation reflected by the mirror is comparable in power with the input radiation. But medium blooming means that in the spectral range of the Doppler profile, which is subjected to the action

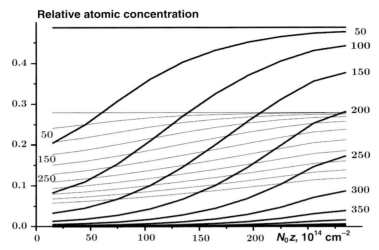

Fig. 4.4 Spatial distribution of a part of ^{64}Zn (bold lines) and ^{66}Zn (thin lines) atoms in a mixture. The curves correspond to a 50 pulse increment. An increase in the number of pulses corresponds to lower atomic concentration of isotopes. The radiation is directed to the greater coordinates and has an average input power of 3 W. The number of pulses passed is given near the lines.

of radiation, the "burnt" isotope is exhausted. Then, the reflected radiation will mainly excite the isotope to be enriched. Hence, the efficiency of isotope selection falls.

In the case of independent counter-propagating radiation beams, the time lapse between pulses may be chosen in such a way that the "hole" in the Doppler profile would have enough time to disappear due to collisions between atoms. An inert buffer gas may be used to increase the rate of this process. Simple estimates show

Table 4.1 Calculation results for separation of zinc isotopes in a chamber with transversal gas circulation and unidirectional radiation beam

Radiation power	3 W
Pulse repetition frequency	10 kHz
Half-height width of the radiation line	500 MHz
Transversal cross-section area of irradiated zone	4.5 cm^2
Parameter $N(0)L$	3×10^{16} cm^{-2}
Contamination of ^{66}Zn isotope by isotopes ^{64}Zn and ^{67}Zn	0.0099
Number of pulses during the transit time	500
Number of quanta per single ^{66}Zn atom	20.0
Absorption in active zone	52%
Yield of ^{66}Zn (output to input flux ratio)	31%
Productivity	2.3×10^{17} atom s^{-1}

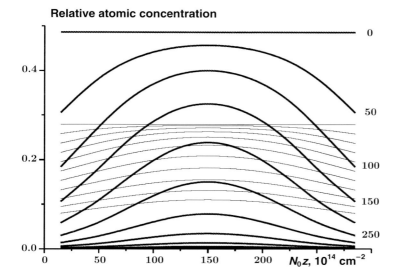

Fig. 4.5 Spatial distribution for ^{64}Zn (bold lines) and ^{66}Zn (thin lines) atoms with counter-propagating beams. The curves correspond to increments of 50 pulses. An increase in the number of pulses corresponds to lower atomic concentration of isotopes. The radiation is directed to larger coordinates and has an average input power of 3 W. The number of passed pulses is given near the lines.

that at the repetition frequency of laser pulses 10 kHz, the buffer gas concentration of $(2–4) \times 10^{14}$ cm^{-3} would be sufficient.

The distributions for ^{64}Zn and ^{66}Zn isotopes at the same geometrical and concentration parameters as those given in Fig. 4.4 are shown in Fig. 4.5, however, for counter-propagating beams. The repetition frequency of radiation pulses in each direction was 10 kHz. The corresponding calculation results are given in Table 4.2.

Tables 4.1 and 4.2 show that at lower radiation power the productivity of isotope separation is greater in the case of counter-propagating beams as compared to the case of unidirectional radiation. It was shown above that the calculation results

Table 4.2 Calculation results for separation of zinc isotopes in a chamber with transversal gas circulation and counter-propagating radiation beams

Total radiation power in both directions	2.3 W
Contamination of ^{66}Zn isotope by isotopes ^{64}Zn and ^{67}Zn	0.01
Number of pulses during the transit time	500
Number of quanta per single ^{66}Zn atom	13.7
Absorption in active zone	62 %
Yield of ^{66}Zn (output to input flux ratio)	35 %
Productivity	2.6×10^{17} atom s^{-1}

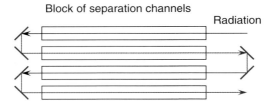

Fig. 4.6 Multimodule separation unit.

depend on the parameter $N(0)L$ (product of atomic concentration at the input of the flux and the length of active medium). At high concentrations, the resonance transfer of excitation and exchange chemical reactions are possible. They reduce the efficiency of isotope separation. Consequently, the atomic concentration should be reduced simultaneously with an increase in the active medium length. The specific values of atomic concentrations depend on the rate constants of the reactions mentioned above and on the desired degree of isotope separation. At the concentration of 10^{13} cm^{-3}, the length of the optical path in the calculations under discussion is 30 m. Under these conditions, it is desirable to use a multimodule construction of the separation unit (see Fig. 4.6).

4.3.2
Longitudinal Gas Circulation

Let us consider a variant of longitudinal gas circulation at the radiation power of 2 W. The distribution of the two isotopes along the tube axis is shown in Fig. 4.7. At the output ($z = 0$), the concentration of ^{64}Zn isotope is considerably reduced

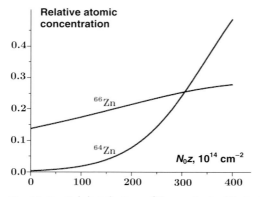

Fig. 4.7 Spatial distribution of the content of Zn isotopes at longitudinal gas circulation. The output radiation power is 2 W. Radiation is directed toward the smaller coordinate.

(lower by almost two orders of magnitude) as compared to the initial mixture. On the other hand, the amount of ^{66}Zn was reduced by two times only, hence, a considerable enrichment of this isotope is obtained. The general characteristics of this mode of operation are given in Table 4.3.

Table 4.3 Calculation results for separation of zinc isotopes in a chamber with longitudinal gas circulation

Radiation power	2 W
Contamination of ^{66}Zn isotope by isotopes ^{64}Zn and ^{67}Zn	0.029
Number of pulses during the transit time	3100
Number of quanta per single ^{66}Zn atom	6.75
Absorption in active zone	71 %
Yield of ^{66}Zn (output to input flux ratio)	49 %
Productivity	4.6×10^{17} atom s^{-1}

The characteristics of the method strongly depend on detuning of the radiation frequency (see Fig. 4.10). A strong minimum is observed in the contamination of ^{66}Zn by undesirable isotopes. This is explained by the fact that at small detuning the excitation selectivity becomes worse, whereas at large detuning the absorption of radiation falls.

The flux velocity at a fixed number of pulses in a time lapse required for radiation to pass through the active zone, other initial parameters being the same, is proportional to the length of the zone. It equals to approximately 3 m s^{-1} at a length of active zone of 1 m. But in long-term media it may be greater, up to a few dozens meters per second. In this case the isotope separation with longitudinal vapor circulation is inferior to the scheme with the transversal circulation.

In this section we have shown, with an example of such inconvenient element for isotope separation as zinc that the method suggested possesses acceptable energy efficiency and competes with the other methods.

4.4
Output Parameters Versus the Detuning of Radiation Frequency

From the practical point of view, it is important to understand how sensitive are the output characteristics to variations of operating parameters. Let us take a set of input parameters corresponding to Table 4.1 as the initial variant. First, let us consider the sensitivity of output parameters to radiation frequency detuning. This parameter can be easily varied and thus we can control the process.

The coordinate dependencies of the concentrations of ^{64}Zn and ^{66}Zn isotopes are shown in Figs. 4.8 and 4.9 at various detunings. The total input atomic concentration was 5×10^{14} cm^{-3} and the length of the active zone was 60 cm. The parameter $N(0)L$, similar to the previous calculations, was 3×10^{18} cm^{-2}. One can

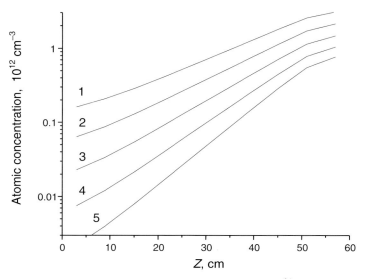

Fig. 4.8 Coordinate dependence of the concentration of ^{64}Zn at various detuning relative to the peak value: (1) 2.2 GHz; (2) 2.15 GHz; (3) 2.10 GHz; (4) 2.05 GHz; (5) 2.0 GHz. The radiation is directed toward the lower coordinate.

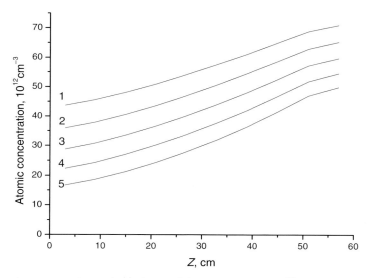

Fig. 4.9 Coordinate dependence of the concentration of ^{66}Zn at various detuning relative to the peak value: (1) 2.2 GHz; (2) 2.15 GHz; (3) 2.10 GHz; (4) 2.05 GHz; (5) 2.0 GHz. The radiation is directed toward the lower coordinate.

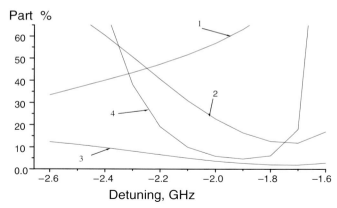

Fig. 4.10 Principal output characteristics versus frequency detuning: (1) the total absorption; (2) ratio of output flux of ^{66}Zn to its input flux; (3) yield of ^{66}Zn atoms per single quantum of radiation; (4) degree of contamination (the contamination of ^{66}Zn by ^{64}Zn and ^{67}Zn isotopes is implied).

see that the concentration of ^{64}Zn depends strongly on the coordinate, whereas the corresponding dependence for ^{66}Zn isotope is relatively weak. It is worth noting that a slight change in frequency detuning by 0.1 GHz leads to a considerable change in the coordinate distribution for ^{64}Zn, and, as a consequence, to a change in material purity.

The dependence of the main output characteristics on frequency detuning is shown in Fig. 4.10. Detuning influences most strongly the contamination of material by ^{64}Zn and ^{67}Zn isotopes. One can see that the degree of contamination has a noticeable minimum (0.005) at the detuning of 1.9 GHz. Nevertheless, at greater (by absolute value) detuning, the effective yield and quantum efficiency increase for ^{66}Zn. It makes sense to shift to greater detunings. This also reduces the risk of falling into the abruptly rising right branch of the contamination curve. In changing the detuning from 2.2 GHz to 2.0 GHz, the degree of contamination falls by almost three times. But in this case, the output of the selected isotope considerably reduces and the number of quanta per single ^{66}Zn atom increases. Absorption, as expected, rises at smaller detuning.

The distribution of absorption along the atomic flux is shown in Fig. 4.11. The medium is almost opaque at the input, whereas the absorption at the output is negligible. The integral absorption over the cross-section is close to 0.5.

The results presented show that favorable conditions for isotope separation exist in a sufficiently narrow range of detuning. A criterion for the proper detuning choice may be, seemingly, the value of the integral absorption. In the conditions under consideration, it should be in the interval 0.5–0.6.

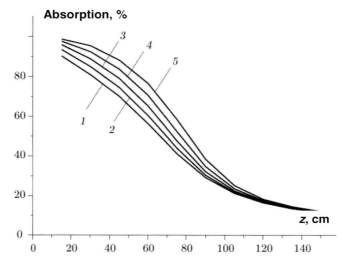

Fig. 4.11 Absorption of radiation versus the coordinate along the flux of atoms. Curves correspond to the detunings: (1) 2.2 GHz; (2) 2.15 GHz; (3) 2.10 GHz; (4) 2.05 GHz; (5) 2.0 GHz.

4.5
Influence of the Radiation Line Profile on Output Characteristics of the Separation Process

The mathematical model allows for a radiation line profile. The choice of the proper profile shape closest to the real profile is rather difficult. An exact experimental registration of the profile with sufficient resolution is possible, however, with considerable difficulties. A typical profile consists of a few peaks corresponding to different modes. The number of modes and the profile width depend on a particular construction and there are methods for mode selection. The profile width varies from 200 MHz (single-frequency mode, in which a dye laser is used as a generator in an amplifying line) to 2 GHz. Experimentally measured profiles have complicated shape and vary from pulse to pulse. It is reported for the so-called profile substrate whose width may reach 5–6 GHz. Needless to say that measures should be taken in the considered case for eliminating the substrate. The width of a separate peak in a profile is about 200 MHz, which may be comparable with the separation between the peaks. The width is firstly determined by the optical properties of the cavity with a diffraction grating and by those of active medium. Under the considered conditions, the field broadening ($\sim$ 30–50 MHz) and the broadening due to the finite duration of the pulse ($\sim$ 50–80 MHz) have negligible influence.

Having no exact profile, let us compare calculation results for three kinds of profile shapes: rectangular, Gaussian, and Lorentz assuming that their peaks are completely overlapped. Let us consider, as we did above, the action of single-frequency radiation shifted to lower frequencies from the peak corresponding to ^{64}Zn. The degree of contamination of the separated isotope will be determined relative to the concentration of ^{64}Zn and ^{68}Zn. The allowance made for the second radiation frequency would not result in principally new effects because its role mainly reduces to the excitation of a relatively small number of ^{68}Zn atoms.

The pumping rate (volume of vapor pumped in unit time) will be calculated via the number of the excitation pulses N in a time lapse needed for passing the active zone. The relationship between these parameters is as follows:

Pumping rate = Pulse frequency × Active volume/N

Let us compare the results of calculating various profiles for transversal circulation with the initial data taken from Table 4.1. In Table 4.4, the parameters for various profiles of the radiation line are given. One can see that the Lorentz profile results in about four times greater degree of contamination as compared to the others. A positive feature of the rectangular profile is the 1.5-fold higher quantum efficiency and, hence, productivity. The Gaussian and Lorentz profiles differ significantly only in the degree of contamination. The slightly greater absorption coefficient in the case of the Lorentz profile is explained by ^{68}Zn absorption. It is worth noting that data presented in Table 4.4 correspond to the conditions that are optimal for the rectangular profile; they differ from the optimal conditions in other cases.

Table 4.4 Calculation parameters for three different profiles of the radiation line

Parameters	Rectangular profile	Gaussian profile	Lorentz profile
Contamination of ^{66}Zn isotope by isotopes ^{64}Zn and ^{67}Zn	0.0099	0.0088	0.038
Number of selected atoms per radiation quantum	0.050	0.036	0.036
Absorption in active zone	52%	55%	61%
Yield of ^{66}Zn (output to input flux ratio)	31%	22%	22%

The dependence of the output characteristics on the half-height width for various profiles is shown in Fig. 4.12. We are interested in those initial parameters and output characteristics which provide low contamination (1%). Among these initial parameters, the pulse energy, the concentration of zinc atoms, and the rate of circulation can be easily varied. In the calculations, the atomic concentration was fixed and the number of pulses corresponding to the required degree of contamination was determined. All the curves correspond to 1% contamination of the output flux averaged over the tube length.

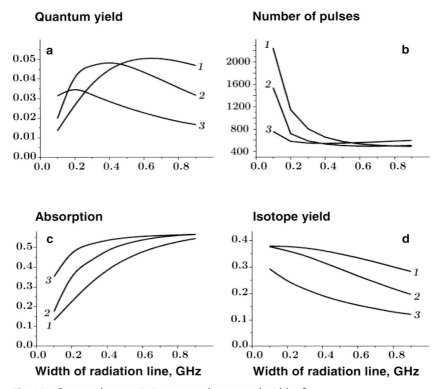

Fig. 4.12 Output characteristics versus the spectral width of radiation line: (1) the rectangular profile of radiation spectrum; (2) Gaussian profile; (3) Lorentz profile.

The most important characteristic is the quantum yield of ^{66}Zn, that is, the number of isotope atoms of prescribed purity per unit radiation quantum. System productivity (atoms s^{-1}) is the product of quantum yield and the radiation power expressed in the number of quanta emitted for 1 s. The quantum output versus the radiation line width for different spectrum profiles is shown in Fig. 4.12a. In all three cases, a maximum of quantum yield is observed. For the rectangular and Gaussian profiles it is about 5 % and for Lorentz profile it is 3.5 %. The considerable difference indicates a great influence of far wings in the radiation spectrum. Efforts on removing them is undoubtedly warranted. A positive aspect is that the maximum quantum yield itself is relatively flat and there is no need to precisely prescribe the line width. For Lorentz profile, the maximum quantum yield corresponds to the line width of approximately 200 MHz, which agrees with the single-frequency mode of generation. For other profiles smoothed over modes, the generation may occupy several modes. In these cases, the flatness of the quantum yield dependence on line width gives the hope that in multimode operation the quantum yield would not fall noticeably as compared to the equivalent smoothed profile.

The dependence of the number of pulses versus the radiation line width is shown in Fig. 4.12b for the considered profiles. Curves a and b show that the optimal quantum yield in all the cases is obtained at approximately the same number of pulses ($\sim$500). Here, we have to make an important note. Calculations show that the number of pulses or the rate of circulation should be determined with high accuracy of the order of a few percent. A minor excess of the number of pulses leads to a marked reduction in quantum yield because the last pulses excite the isotope to be separated.

For all three profiles considered, the maximal quantum yield occurs at approximately the same absorption ($\sim$50 %), which is seen in Fig. 4.12c. The absorption can be easily measured and may be a criterion of the process efficiency. Establishing such a criterion is one of the main problems in theoretical investigation.

The parameter "yield of the selected isotope" is determined as the ratio of the input flux of ^{66}Zn atoms to the output flux, and characterizes the part of atoms that can be extracted as a useful product. In the considered case, at optimal conditions it is close to 30 % (see Fig. 4.12d). Most (70 %) of the ^{66}Zn atoms are excited by radiation and extracted with chemical compounds. The yield of the selected isotope can only be increased at the sacrifice of worsening the purity of the material obtained.

With 50 % absorption and 30 % yield of the isotope to be selected, it is possible to determine the general balance relationship between the input atomic flux and the flux of quanta under optimal conditions. Taking into account that 70 % of the ^{66}Zn atoms and almost all the atoms of the other isotopes are excited and allowing for 50 % absorption we obtain

Quantum flux = 1.85 × atomic flux

The above relationship is approximately the same for various profiles of the radiation spectra.

Hence, employment of the Lorentz profile for radiation spectrum worsens noticeably the productivity of the system as compared to the rectangular or Gaussian profile. Consequently, it is reasonable to "cut" the profile wings. If the profile wings are of the Lorentz type or are even more pronounced, then the single-frequency radiation should be used.

The width of the Gaussian or the rectangular profile may vary within 150 MHz relative to the optimal profile width.

The balance ratio between the quantum flux and the input flux of Zn atoms is almost the same for all the three profiles considered.

In all the cases, isotope separation is quite efficient (as compared to other methods).

4.6
Experiments on Laser Separation of Zn Isotopes by the Photochemical Method

In [122], the experimental laser separation of Zn atoms by the single-photon method was reported. The corresponding experimental setup is schematically shown in Fig. 4.13. The separation zone is the quartz tube 60 cm in length and 3 cm in diameter heated by an external oven. The flux of the gas mixture including the argon atoms under consideration, and the gas-reagent molecules propagates through the tube. The circulation rate of the mixture was $\geq 0.5\,\mathrm{l\cdot s^{-1}}$, argon pressure was 1–2 Torr, and the concentration of the gas-reagent is $10^{16}\,\mathrm{cm^{-3}}$. Under these conditions the drift time of the atoms through the interaction zone is about 1 s. Atoms of the selected isotope pass to the separation zone from a specially heated reservoir at the temperature that provides the required concentration. Evacuation is made by a roughing-down pump. The radiation of the power laser propagates along the flux providing "burning" of the excited atoms due to chemical reaction. The products of this process deposit on the walls of the separation chamber. Vapors of the other isotopes are condensed on the walls of a collector chamber at room temperature. The chemical reaction is provided by collisions of Zn atoms in $(4p^3P_1)$-state excited by laser radiation (see Fig. 4.14) with $(C_2H_5)_2O$ molecules. The rate constant of this reaction is $k^* = 1.61 \times 10^{-9}\,\mathrm{cm^3\,s^{-1}}$. The gas-reagent was chosen from a set of molecules to provide a large value of the rate constant, and selectivity of isotope separation in agreement with the conditions formulated in Chapter 3. The maximal value of the rate constant for unexcited atoms is $10^{14}\,\mathrm{cm^3\,s^{-1}}$, which is five orders in magnitude less than that for the $(4p^3P_1^0)$ state. This value was determined experimentally by detecting

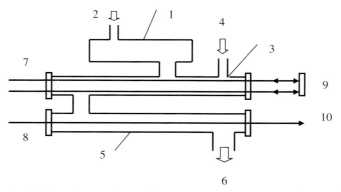

Fig. 4.13 Experimental setup for separating Zn isotopes by the single-photon method: (1) reservoir with Zn; (2) buffer gas feeding; (3) separation chamber; (4) gas-reagent feeding; (5) collector chamber; (6) evacuation of gas mixture; (7) power laser radiation; (8) probing laser radiation; (9) mirror; (10) photodetector.

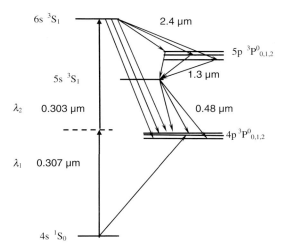

Fig. 4.14 Working transitions of Zn atoms.

the concentration of Zn atoms in the collector chamber with and without the gas-reagent.

Selective, with respect to atomic velocities, excitation of the absorption lines nonuniformly broadened by narrow-band laser radiation is used for the elements whose isotopic shift is overlapped by Doppler broadening. In this case, only atoms with a definite projection of the velocity v in the direction of wave propagation interact with the light field [182]:

$$|v_0 - v + v_0 v/c| \leq \gamma, \quad \Delta v \qquad (4.16)$$

where v_0 is the central frequency of the atomic absorption line; v and Δv are the frequency and the width of the radiation line; γ is the uniform width of the atomic absorption line. The cross-section of the radiation of frequency v that induces the transition from level 1 to level 2 is defined by the formula [18]:

$$\sigma(v) = \sigma(v_0) a(v)/a(v_0) \qquad (4.17)$$

Here, σ_0 is the absorption cross-section at the central transition frequency of the Doppler profile.

$$\sigma(v_0) = \frac{c^2 g_2 a(v_0)}{8\pi v_0^2 \tau_{21} g_1} \qquad (4.18)$$

where $1/\tau_{21} = A_{21}$ is the probability of spontaneous transition from level 2 to level 1, g_2 and g_1 are the statistic weights of the upper and lower states, respectively. Function $a(v)$ is the normalized profile of the absorption line. In the case of Doppler broadening, it is defined by the expression

$$a(\nu) = \frac{2\sqrt{\ln 2}}{\sqrt{\pi}\Delta\nu_D} \exp\left(-\frac{4(\ln 2)(\nu - \nu_0)^2}{(\Delta\nu_D)^2}\right) \quad (4.19)$$

where $\Delta\nu_D$ is the half-intensity Doppler width of the line. If the gas temperature is T and the atomic mass is M, then we have

$$\Delta\nu_D = 2\nu_0\sqrt{\frac{2kT\ln 2}{Mc^2}} \quad (4.20)$$

The absorption coefficient for resonance laser radiation propagating in a gas medium with concentration n is $\alpha(\nu) = \sigma(\nu)n$. Hence, the frequency dependence of the absorption coefficient is as follows:

$$\alpha(\nu) = \alpha_0 \exp[-4(\ln 2)((\nu - \nu_0)^2/(\Delta\nu_D)^2)] \quad (4.21)$$

where α_0 is the absorption coefficient at the center of line. For an isotopic mixture, the common profile of the absorption line consists of the sum of coefficients (4.21) with the corresponding values of ν_0 and α_0 that are proportional to their relative content.

The frequency dependencies of the absorption calculated for the transition $4s^2\,^1S_0 \to 4p\,^3P_1^0$ (see Fig. 4.14) at $T = 350\,°C$ are shown in Fig. 4.15 for the

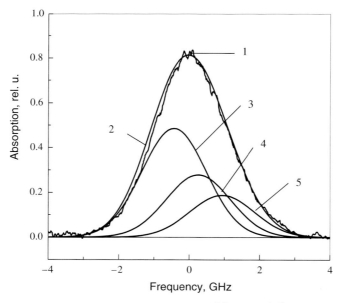

Fig. 4.15 Absorption spectrum of Zn for $4s^2\,^1S_0 \to 4p\,^3P_1^0$ transition: (1) experiment; (2) calculation by formula (4.21); (3–5) calculation for ^{64}Zn, ^{66}Zn, and ^{68}Zn isotopes, respectively.

4.6 Experiments on Laser Separation of Zn Isotopes by the Photochemical Method

main Zn isotopes along with the corresponding total profile. One can see that the absorption spectra of various isotopes in the gas flux strongly overlap and no selectivity can be spoken of. Nevertheless, as it was mentioned at the beginning of the chapter, if the frequency of the exciting radiation is shifted to one side of the total absorption profile, then the absorption coefficients would strongly differ for different isotopes. For the case of ^{64}Zn and ^{66}Zn isotopes, this is illustrated in Fig. 4.16. The ratio of the absorption coefficients is greater than ten at the detuning of 2 GHz. The total absorption coefficient also falls noticeably in this case, however, the long time of interaction between the radiation and the atoms provides 100 % efficiency of exciting atoms of desired isotope at the corresponding experimental parameters (see below). The other isotopes are also excited, however, their participation is rather small. In addition to the selected isotope, they also participate in photochemical reaction and are "burnt out" in the interaction chamber, which lowers the separation efficiency marginally.

In the experiments, Zn atoms were excited at the intercombination transition $4s^{2\,1}S_0 \rightarrow 4p^3P_1^0$ ($\lambda = 307$ nm) (see Fig. 4.14) in the separation zone along the flux from the source of narrow-band tunable pulsed laser radiation (power laser) described in [50] and in Chapter 2. The average power of the laser radiation propagating in the separation zone was $\sim$2 W. The beam diameter was $\sim$1 cm at the repetition frequency of 12 kHz, the pulse duration of 10 ns, and the width of the radiation line $\Delta \nu = 36$ MHz. Since the total absorption per single passage of laser radiation was small (10–20 %), for better performance it was returned back by a mirror to the separation chamber (see Fig. 4.13). In this variant, the upper excited state has a relatively long lifetime ($\tau = 10.5$ µs). This value was obtained in

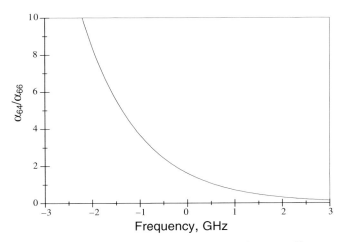

Fig. 4.16 Ratio of the absorption coefficients for ^{64}Zn and ^{66}Zn isotopes versus the frequency detuning from the center of the total absorption profile.

real experimental conditions by measuring the decay time of luminescence from the excited levels without gas-reagent in the separation zone (see Chapter 3). In Fig. 4.15, the calculation results and the experimental absorption spectrum taken at the same temperature are shown. The calculated value of the total absorption profile agrees well with the experimental data.

If the pulse duration of the resonance laser radiation is much shorter than the lifetime of the upper state, then the probability of exciting an atom during the pulse action is determined by the expression

$$W(\nu) = \sigma(\nu)\Phi/S \tag{4.22}$$

where Φ is the total number of photons in the pulse and S is the area of the laser beam cross-section. In the experimental conditions under consideration ($\nu_0 = 10^{15}$ MHz, $\tau_{21} \approx 10^{-5}$ s, $g_2 = 3$, $g_1 = 1$, $\Phi \approx 2.5 \times 10^{14}$ photons), the estimate by formula (4.22) yields $\sigma(\nu_0) \approx 10^{-14}$ cm^2 and $W(\nu_0) \approx 2.8$. Hence, in a time lapse of pulse duration almost all the atoms are excited in the interaction with laser radiation. The frequency shift from the center of the absorption profile by -2 GHz reduces the excitation probability ($W \approx 4.5 \times 10^{-2}$). Nevertheless, atoms interact with radiation during all the time of their drift in the gas flux (≈ 1 s). In this time, approximately 10^4 radiation pulses pass through the working medium and despite a low probability of excitation by a single pulse, almost all the atoms transfer to the upper excited state during the drift process.

During the action of the excited pulse, only a small part of the atoms whose velocities directed along the direction of radiation satisfy condition (4.16) is excited. This value is doubled due to the radiation reflected by a return mirror and propagating in the reverse direction. Since the time lapse between impacts is short ($\sim 10^{-7}$ s), a great number of impacts occur in the time lapse between pulses ($\approx 10^{-4}$ s), thus restoring Maxwellian distribution and all needed atoms participate in the separation process.

The isotope composition of the separated products passed into the collector chamber was analyzed by means of probing radiation of another laser source with the radiation frequency tuned over the whole Doppler absorption profile of the same transition. The laser intensity was three orders of magnitude lower as compared to that of the power laser for excluding the influence of chemical reactions that occur in the collector chamber. By detecting variations of the Doppler absorption profile in the collector chamber, with or without the power radiation in the interaction chamber, it is possible to observe changes in isotope composition in the collector chamber. The calculated dependence of the shift of the maximum Doppler absorption profile for zinc atoms versus the content of ^{64}Zn isotope in the mixture is shown in Fig. 4.17. Hence, by fixing the absolute values of the position of the absorption maximum it is possible to determine the isotopic composition of the product. One more method used in the experiments under discussion is

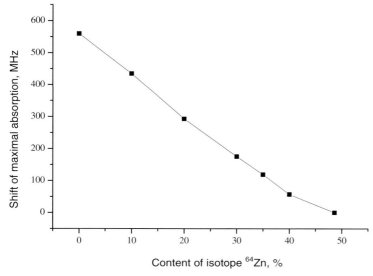

Fig. 4.17 Shift of the maximum of the Doppler absorption profile of zinc atoms versus the content of ^{64}Zn isotope in a mixture.

the direct mass-spectrometric analysis of the product deposited on the walls of the collector chamber.

If the frequency of the power laser radiation is adjusted to the center of the Doppler profile for zinc atoms, then in the separation chamber almost 100 % "burning" of atoms is observed. This fact proves the efficiency of exciting zinc atoms and of the chemical reaction

$$\text{Zn}^* + (\text{C}_2\text{H}_5)_2\text{O} \rightarrow \text{ZnO} + (\text{C}_2\text{H}_5)_2 \tag{4.23}$$

The product ZnO in reaction (4.23) is deposited on the walls of the interaction chamber. Sufficiently efficient chemical reactions are observed with some other molecules, however, the best selectivity is obtained with diethyl ether molecules $(\text{C}_2\text{H}_5)_2\text{O}$. This fact may be explained by the weaker influence of various secondary reactions discussed in Chapter 3.

At a higher degree of burning, the problem of insufficient quantity of separated isotopes in the reaction zone arises. Indeed, in a time lapse between the pulses ($\approx 10^{-4}$ s) the atoms may not have enough time to pass into the interaction zone during the diffusion period ($t_{\text{diff}} = r^2/D$). For example, at the characteristic dimensions $r \approx 1$ cm and average values of diffusion coefficient at normal pressure $D \sim 0.1$ cm^2 s^{-1}, the diffusion time under the experimental conditions considered is very long $t_{\text{diff}} \sim 10^{-2}$ s. All these speculations are confirmed by experiments in which the average power of the laser radiation is weakened by a

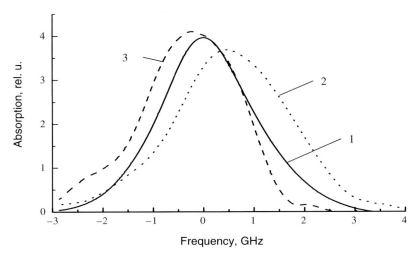

Fig. 4.18 Absorption spectrum of Zn in a collector chamber corresponding to $4s^2\ 1P_0^1$ transition: (1) Doppler profile without power laser; (2) the power laser is adjusted to -2 GHz relative to the center (the values are multiplied by a factor of 8); (3) the power laser is adjusted to $+2$ GHz relative to the center (the values are multiplied by a factor of 20).

mechanical interrupter while keeping the pulsed power unchanged. Under these conditions, the absorption of radiation in the interaction zone increases.

If the frequency of laser radiation is shifted from the center of the Doppler profile, then the profile of the absorption line deforms, which may be detected in the collector chamber. The corresponding experimental results are shown in Fig. 4.18 for the atomic concentration in the separation zone $\approx 10^{13}$ cm^{-3}. If the frequency shift is -2 GHz, then the maximum of the Doppler profile moves to the right. This is connected with the preferential excitation followed by "burning out" of ^{64}Zn isotopes. The opposite situation is observed for the frequency shift $+2$ GHz. One can see in Fig. 4.18 that the number of "burnt out" atoms detected in the collector chamber exceeds the number of atoms excited in the separation chamber. This, seemingly, may be connected with secondary chemical reactions including radicals produced and unexcited atoms. The comparison of the shift of the maximum of Doppler profile with the calculated values (see Fig. 4.17) shows that in the separation chamber more than 95 % of ^{64}Zn isotope is "burnt out." Similar results are obtained from mass spectrometric analysis of zinc deposited on the walls of the collector chamber.

The product yield obtained was ~ 1 g at a 3-hour exposition under the action of UV radiation of average power 2 W at the circulation rate of mixture 100 l s^{-1}. The total cost of the product is \$50 per gram. The advantage of the method described, as compared to the two-photon method, which will be discussed in Chapter 5, and

4.7 Experiments on Laser Separation of Rubidium Isotopes by the Photochemical Method

Laser separation of rubidium isotopes by ordinary AVLIS methods encounters certain problems. First, there is no efficient scheme of photoionization with presently available tunable laser sources. Second, high saturated vapor pressure at room temperature leads to a loss of selectivity.

In [122], the single-photon method for laser separation of rubidium isotope based on "burning" selectively excited Rydberg ($11P_{3/2}$)-states of Rb atoms (see Fig. 4.19) due to chemical reaction was experimentally studied. A natural mixture of rubidium atoms consists of two odd ^{85}Rb (72.17 %) and ^{87}Rb (27.83 %) isotopes. The corresponding isotopic shift is small, however, the absorption lines of different isotopes are found to shift by a value exceeding the value of Doppler broadening. The oscillation strength of the transition $4S_{1/2} \rightarrow 11P_{3/2}$ is not large, hence a high power laser radiation density is needed for saturation. The absorption cross-section (4.17) can be calculated by the formula

$$\sigma(\nu)[cm^2] = 2.7 \times 10^{-2} f a(\nu) \tag{4.24}$$

where ν is in Hz. For the considered transition we have $\lambda = 311$ nm; $f = 7.6 \times 10^{-5}$ [183]; $\sigma(\nu_0) \approx 10^{-15}$ cm^2. The value obtained is even less than the absorption cross-section of the intercombination transition in Zn atom. By using formula (4.22), we find that the probability of exciting ($11P_{3/2}$) state by a single pulse of laser radiation with the energy of 5×10^5 J is $W(\nu_0) \approx 8 \times 10^{-2}$. This is a small value,

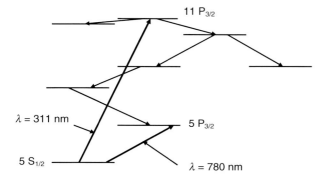

Fig. 4.19 Working transitions in rubidium atom.

however, in exciting in a flux, where the average drift time is ~1 s by laser radiation source operating at high repetition frequency (12 kHz) all "wanted" atoms have time to be excited to the upper (11P$_{3/2}$ state. The selectivity of the process is made increased by employing chemical reaction for removing the excited atoms.

The experimental setup for separating rubidium isotopes is shown in Figs. 3.3 and 4.20. Rb atoms pass from a heated reservoir in argon flux at the pressure of approximately 1 Torr to the interaction chamber, where they are selectively excited. The separation chamber is a glass tube 3 cm in diameter heated by an external oven. The length of the zone in which atoms interact with laser radiation was 80 cm. The rate of circulation maintained at 0.5 l s^{-1}.

The atomic transition 5S$_{1/2}$ →11P$_{3/2}$ (see Fig. 4.19) was excited at wavelength $\lambda = 311$ nm by a source of narrow-band tunable pulsed radiation already mentioned in the separation of zinc atoms (see Section 4.6). The average power of the laser radiation in the interaction zone was ~0.6 W at the beam diameter of 1 cm. The gas-reagent was fed into the same chamber.

For analyzing the isotope composition of rubidium atoms that were excited and which reacted with the gas-reagent, radiation of a frequency-tunable diode laser operating near the frequency of the transition 5S$_{1/2}$ →5P$_{3/2}$ (D$_2$-line of rubidium, $\lambda = 780$ nm) was propagated in the transversal direction at the end of the interaction chamber. The spectral width of the radiation was less than 60 MHz, which provided a high spectral resolution of hyperfine structure.

The absorption spectra of the rubidium atoms in the interaction chamber corresponding to the transition 5S$_{1/2}$ →11P$_{3/2}$ detected by the absorption of radiation and by the luminescence that arises due to decay of the excited state are shown in

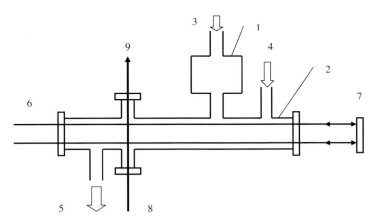

Fig. 4.20 Experimental setup for separating rubidium isotopes by the single-photon method: (1) reservoir with Rb atoms; (2) interaction chamber; (3) Ar supply; (4) input of gas-reagent; (5) evacuation; (6) laser radiation; (7) return mirror; (8) input of diode laser radiation; (9) photodetector.

4.7 Experiments on Laser Separation of Rubidium Isotopes by the Photochemical Method

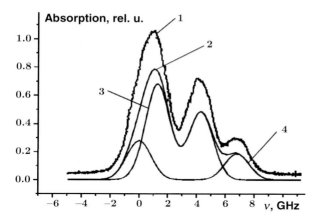

Fig. 4.21 Absorption spectrum of rubidium atoms corresponding to the transition $5S_{1/2} \rightarrow 11P_{3/2}$: (1) experimental results; (2) calculated data; (3) calculated profile for ^{85}Rb; (4) calculated profile for ^{87}Rb.

Fig. 4.21. Both methods gave the same spectra. At argon pressure of up to ∼5 Torr and concentration of rubidium atoms ∼10^{13} cm^{-3} no pronounced broadening of the absorption spectra was observed. Hence, the selectivity of excitation persists. Also, the absorption spectra calculated at the temperature of 120 °C are presented in Fig. 4.21. A good agreement is observed between the experimental and calculated values. The calculated positions of the hyperfine structure components of ^{85}Rb and ^{87}Rb isotopes are shown in the same figure. One can see that employment of the last two peaks for excitation yields rather good selectivity for one of the isotopes.

A considerably high concentration of buffer gas (up to 10^{17} cm^{-3}) and the pulsed mode of the excitation source used result in that during the time lapse between pulses a great number of atom impacts with other particles occur (in the considered conditions ∼10^3). This promotes a complete mixing of hyperfine structure components for sodium atom [184] in collisions of the latter with He and Ar atoms. The absolute characteristics of such a process are an order of magnitude better than gas-kinetic values. Hence, there is no need for using two-frequency laser radiation for exciting all the atoms in such cases.

The lifetime of an excited ($11P_{3/2}$)-state was measured by visible and infrared luminescence that arises due to the cascade decay of the latter. For selective removal of the excited atoms from the flux, a photochemical reaction with methanol or diethyl ether was used. After adding the gas-reagent to the interaction chamber, the lifetime of the excited state ($\tau_0 \approx 0.55$ µs) decreased noticeably. By measuring the decay time of the luminescence versus pressure, the rate constants were found for the chemical reactions of rubidium atoms in ($11P_{3/2}$)-state with methanol ($k = 1.47 \times 10^{-9}$ cm^3 s^{-1}) and diethyl ether ($k = 8.4 \times 10^{-10}$ cm^3 s^{-1})

molecules. It is sufficient for the "burning out" of selectively excited atoms during their lifetime. The products of these reactions are, seemingly, RbO and RbOH that are well accommodated at the walls of the reaction chamber. Rubidium atoms excited to ($11P_{3/2}$) state efficiently react with some other molecules. This was detected by the fall in the atomic concentration at the end of the interaction zone. Nevertheless, these are the molecules mentioned above that provide the most efficient selectivity of the process.

In Fig. 4.22, the absorption spectra of rubidium atoms are shown that were detected at the end of the interaction zone by the absorption of CW diode laser radiation in D_2-line of rubidium. The wall temperature in the interaction chamber was 120 °C and the concentration of rubidium atoms was $\approx 10^{12}$ cm^{-3}. The latter parameter was limited because at the absorbing length of 3 cm, determined by the diameter of the glass tube in the interaction chamber, the signal is completely absorbed in the transition $5S_{1/2} \rightarrow 5P_{3/2}$ at higher concentrations.

The Doppler width of the transition at the considered frequency is approximately three times lower than in the case of the excitation transition. All components of the hyperfine structure are resolved. The isotopic composition of rubidium atoms at the end of the interaction chamber can be inferred from the spectra presented. Experimental results show that if the frequency of the exciting radiation is adjusted to the absorption peak of the ^{87}Rb isotope (peak 4 in Fig. 4.21), then at the end of interaction chamber the concentration of this isotope falls by more than 90 % as compared to the initial state.

By condensing atoms passed from the interaction zone it is possible to acquire the product in weight quantities. The process productivity calculated for the experimentally realized conditions is $\sim 10^{-2}$ g day^{-1}. This parameter can be considerably

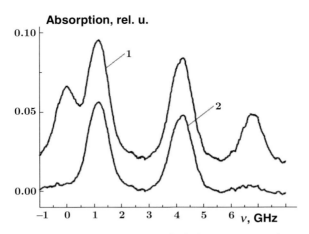

Fig. 4.22 Absorption spectrum of rubidium corresponding to D_2-line: (1) without power laser; (2) with power laser.

improved by increasing the atomic concentration, the rate of gas mixture flux, and the power of the laser radiation. No experiments on producing isotopes by a complete technological cycle were carried out. Nevertheless, since the exponent in expression (4.15) may reach 100, the number of radiation quanta needed for obtaining a single atom of the pure product does not exceed two for ^{85}Rb and six for ^{87}Rb. Now we can calculate the unit cost of the separation: $ 52 per gram for ^{85}Rb and $ 160 per gram for ^{87}Rb.

5
Coherent Isotope-Selective Two-Photon Excitation of Atoms

5.1
Brief Description of Two-Photon Excitation and the Mathematical Model

Isotopes are usually separated via photoionization that occurs in a sequence of incoherent single-photon transitions. An extensive literature is devoted to this method. It is thoroughly considered in the review [19].

The essence of the two-photon excitation of an atom via an intermediate level is that the atom transits to the final state by simultaneously absorbing two oncoming photons with close frequencies and the intermediate level is slightly populated in this process [185]. Practical isotope separation by means of the two-photon excitation became possible due to substantial progress in laser technique in recent years. A tunable UV radiation with an extremely narrow spectral width became available at a high peak and average power (see Chapter 2). The line width became narrower than the characteristic isotope shifts, which made possible efficient selective excitation of atoms of most of the chemical elements.

A general scheme of two-photon excitation of atoms is shown in Fig. 5.1. It is characterized by detuning Δ of the resonance frequency for one of the photons from an intermediate state and by deviation δ of the sum of the radiation frequencies from the exact two-photon resonance (in short, we will call δ *deviation*). The optimal deviation is distinct from zero [62]. For led, zinc, boron, and silicon atoms considered in this book the optimal value of $\delta \ll \Delta$ is in the range of 10–500 MHz at the density of the average radiation power for each transition ~ 2 W cm^{-2} and the repetition frequency ~ 10 kHz. The optimal detuning value is an order of magnitude greater.

Under the conditions of a short-duration pulse radiation (the transverse relaxation times are comparable to or much longer than the pulse duration) and narrow-band radiation spectrum (the spectral width of the radiation line is comparable to the inverse pulse duration), the interaction of radiation with atoms can

---

Laser Isotope Separation in Atomic Vapor. P. A. Bokhan, V. V. Buchanov, N. V. Fateev,
M. M. Kalugin, M. A. Kazaryan, A. M. Prokhorov, D. E. Zakrevskiĭ
Copyright © 2006 WILEY-VCH Verlag GmbH & Co. KGaA, Weinheim
ISBN: 3-527-40621-2

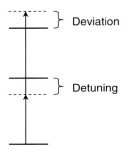

Fig. 5.1 Two-photon excitation of atoms.

be described in the framework of the coherent approximation. According to the theoretical model [63], the dynamics of a three-level scheme can be described by the density matrix approach. In real-number variables the equations for density matrix elements for a three-level atomic system in electromagnetic plane-polarized fields are written as follows [62]:

$$\begin{aligned}
(d/dt)\rho_{11} &= -2\mu_1 V_1 + A_{21}\rho_{22} \\
(d/dt + 1/T_2)\rho_{22} &= 2(\mu_1 V_1 - \mu_2 V_2) + A_{32}\rho_{33} \\
(d/dt + 1/T_2 + W_3)\rho_{33} &= 2\mu_2 V_2 \\
(d/dt + 1/(2T_2))u_1 &= -\Omega_1 \mu_1 - \mu_3 V_2 \\
(d/dt + 1/(2T_2))\mu_1 &= \Omega_1 u_1 + V_1(\rho_{11} - \rho_{22}) + V_2 u_3 \\
(d/dt + 1/(2T_2) + 1/(2T_3) + W_3/2)u_2 &= -\Omega_2 \mu_2 - \mu_3 V_1 \\
(d/dt + 1/(2T_2) + 1/(2T_3) + W_3/2)\mu_2 &= \Omega_2 u_2 + V_2(\rho_{22} - \rho_{33}) - V_1 u_3 \\
(d/dt + 1/(2T_3) + W_3/2)u_3 &= -(\Omega_1 + \Omega_2)\mu_3 + \mu_2 V_1 - \mu_1 V_2 \\
(d/dt + 1/(2T_3) + W_3/2)\mu_3 &= (\Omega_1 + \Omega_2)u_3 - u_2 V_1 + u_1 V_2
\end{aligned} \quad (5.1)$$

where $V_1 = -d_1 E_1/4\pi h$; $V_2 = -d_2 E_2/4\pi h$; E_1 and E_2 are the amplitudes of the electric field intensity for the first and second transitions, respectively; $\Omega_{1,2}$ are the frequency detunings of the transitions; $u_{1,2}$ and $\mu_{1,2}$ are the in-phase and shifted by $\pi/2$ polarization components normalized to the dipole momentum of the transition; u_3 and μ_3 are the real and imaginary parts of element ρ_{13} of the density matrix; $d_{1,2}$ are the dipole momenta of the transitions; ω_{32} and ω_{21} are the frequencies of the transitions; $\omega_{1,2}$ are the radiation frequencies; A_{ij} are the Einstein coefficients; $T_{2,3}$ are the lifetimes of the second and third levels; W_3 is the rate of the upper level depopulation due to external processes (photoionization, superluminescence, chemical reactions). For parallel radiation beams we have $\Omega_1 = \omega_{21}(1 - v/c) - \omega_1$; $\Omega_2 = \omega_{32}(1 - v/c) - \omega_2$, and for oncoming beams we have $\Omega_2 = \omega_{32}(1 + v/c) - \omega_2$, where v is the velocity of the atom. The Maxwellian velocity distribution is assumed for atoms and Doppler broadening for line.

The system of equations (5.1) does not describe the dynamics of an electric field. Hence, the effects related to the deformation of the radiation pulse shape and splitting of the latter into solitons are discarded. Nevertheless, these effects are important if atoms not only absorb radiation but also give up energy to the radiation field [186]. In isotope separation, reradiation of atoms is undesirable because the degree of excitation is reduced and the selectivity falls due to the dynamic Stark effect. Therefore, we will consider radiation pulses of moderate energy, however, sufficient for transferring a considerable number of atoms to the upper level.

5.2
Two-Photon Excitation of Led Atoms

In this section, we consider the possibility of removing the radioactive isotope ^{210}Pb by the photoionization method. Led and led–tin alloys with low radioactivity are necessary in producing microchips with the frequency higher than 500 MHz. Natural led comprises ^{210}Pb isotope, which is a result of the decay of ^{238}U. In a subsequent decay ^{210}Pb transfers to ^{210}Po, which decays into ^{206}Pb through α-decay. Fast α-particles penetrating the (p–n)-junction lead to faults in microchip operation. ^{210}Pb isotope can be removed from led only by isotope separation methods. For laser separation this task is almost ideal [56]. Concentration of this isotope in led is negligible and amounts to $\sim 10^{-13}$ %. In this case, the absorption of radiation by ^{210}Pb atoms is negligible; the problems concerning self-focusing and spatial-temporal matching of radiation pulses are alleviated. Hence, it becomes possible to use lengthy zones for interaction between radiation and substance (of the order of several dozens of meters).

A schematic diagram of led levels that may be used in the photoionization process is shown in Fig. 5.2. The solid lines denote the possible photoionization channels and the dotted lines refer to the transitions related to the spontaneous decay of the levels. The first transition from the ground to the resonance state is determined almost uniquely, because transitions to higher levels require a radiation with too short wavelengths. For the second transition, two variants were considered. In the first variant, radiation of a Nd:YAG laser excites the $7p^3D_1$ level. This type of laser was chosen because the frequency of its radiation is only 6 cm^{-1} above the transition frequency; hence, it can be exactly adjusted to the resonance frequency by varying the temperature of the active crystal. In the second variant, radiation of the dye laser (Rhodamine 6G) excites the $8p^3D_2$ level. The choice of a source of the photoionizing radiation mainly depends on the structure of autoionizing levels and their lifetimes. With the laser complex described above that provides a wide range of tunable radiation, the autoionizing levels were chosen with the cross-section of the order of 10^{-15} cm^{-2}. In this case, an excess of photoionization probability over the probability of radiation decay of the upper level [17] may be provided at a moderate power of ionizing radiation.

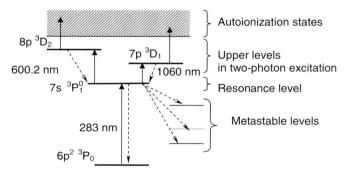

Fig. 5.2 Transitions in the led atom.

The main goal of the calculations is to determine the conditions for maximal productivity of the system taking into account the limitations on the radiation parameters. For the purely radiation problem it reduces to finding the conditions, in which, during a single pulse, in the whole active volume the maximum possible quantity of ^{210}Pb ions would be produced at a prescribed degree of isotope ionization (at least 0.9). For the single transition, the radiation energy is limited by approximately 0.5 mJ.

The choice of radiation parameters in the separation of ^{210}Pb isotope by the photoionization method was discussed in [56]. Calculations were performed under the conditions of an ideally collimated atomic beam. It was found that the principal limitation of system productivity is related to the fact that, in addition to separated ^{210}Pb, whose concentration is negligible in the mixture, radiation is also absorbed by all the remaining isotopes. Minimization of the absorption imposes hard restrictions on the value of the detuning. The latter depends on the radiation energy and pulse duration. For each radiation energy, there exists a particular optimal frequency detuning. At an energy of 50 µJ cm^{-2} the value of the latter is in the range of 30–60 GHz depending on the pulse shape.

In Fig. 5.3, the results of the calculation with the optimal frequency detuning and optimal delay of the ionizing pulse are shown for the transitions $6p^2\,^3P_1 \rightarrow 7s\,^3P_1^0 \rightarrow 8p\,^3D_2$. In addition to the temporal shapes of the pulses, the curves of the relative population of the third (n_3) and metastable (n_m) levels and the ion concentrations (n_i) of ^{210}Pb isotope are presented. The following parameters were used in the considered variant:

- the pulse energy of the first and second transitions was 50 µJ cm^{-2},
- the half-height pulse duration was 15 ns,
- the average probability of photionizing the third level was 10^9 s^{-1},
- the half-height duration of the photoionizing pulse was 10 ns,
- the resonance frequency detuning was 47 GHz,
- the deviation was 30 MHz.

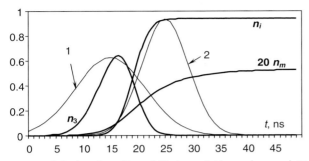

Fig. 5.3 Calculated profiles of (1) the radiation pulses and (2) photoionizing pulses for the transitions $6p^2\,^3P_1 \rightarrow 7s\,^3P_1^0 \rightarrow 8p\,^3D_2$; temporal dependence of the population of the upper (n_3) and metastable (n_m) levels and the ion concentration (n_i).

One can see from Fig. 5.3 that under the conditions described above, the degree of ^{210}Pb isotope ionization exceeds 0.9. The ionization degree of the remaining isotopes is close to 0.001.

In recent years, in the market of materials employed in microelectronics the demand increased for led with ^{210}Pb isotope content lower by a factor of 100–1000 relative to its content in natural led. For obtaining such a material, repeated photoionization is necessary. Led atoms in a collimated beam pass the distance of 10 cm in a time lapse sufficient for triple photoionization to occur at the pulse repetition frequency of 10 kHz. Unfortunately, the third pulse has no effect because a single pulse produces more than 2.5 % of atoms at metastable levels (see Fig. 5.3), which do not participate in photoionization. Special measures are desirable on destructing metastable atomic states in the time interval between the radiation pulses.

In using 50 % power on the first transition, the limiting productivity of the system under the conditions of laser complex discussed is $\sim 1\,\text{g}\,\text{s}^{-1}$ (about 2000 kg of isotopically pure product per month at around-the-clock operation) with the prime cost $\sim \$40\,\text{kg}^{-1}$. Mass production of low-radioactive led by this technology [56] is commercially profitable.

5.3
Two-Photon Excitation of Boron and Silica Atoms

Successful separation of Pb isotopes on a semicommercial scale stimulates study of possible employment of this method for separating other isotopes. In this section, we consider isotopically selective photoionization of boron and silica atoms by means of the frequency-tunable laser radiation.

Natural silica has the following isotope composition: ^{28}Si—92.2 %; ^{29}Si—4.7 %; ^{30}Si—3.1 %. In some technological applications, mainly high isotopic purity of silica is required. Separation of ^{28}Si isotope, whose content is maximal in natural materials, is preferable. The natural content of stable boron isotopes is 19.9 % of ^{10}B and 80.1 % of ^{11}B. The problem is to select ^{10}B from a mixture of the two isotopes for employing it in nuclear reactor industry and to obtain a pure boron isotope for reducing the width of the (p–n)-junction in semiconductors.

For boron and silica atoms, similar photoionization schemes using the two-photon excitation of a highly lying level via a resonance level are suggested. It is essential that the two-photon excitation is implied rather than the two-step excitation. For silica, the scheme of the transitions $3p^2\,^3P_2 \rightarrow 4s\,^3P_2 \rightarrow 5p\,^3D_2$ is considered and for boron it is $2p^2\,^2P_{1/2} \rightarrow 3s\,^2S_{1/2} \rightarrow 4p\,^3P_{1/2}$. The wavelengths of the mentioned transitions for silica and boron coincide with an accuracy of 1.5 %. The oscillator strengths and lifetimes are also close. Hence, the corresponding optimal values of energy densities would be almost equal.

The schematic diagrams of boron and silica atomic levels suitable for using in the photoionization process are shown in Fig. 5.4.

The choice of the first lowest level for boron is not unique. The ground state is doubled with the separation 15 cm^{-1} between the levels. The energy spacing between these levels is far less than the thermal energy of atoms; so we can assume that they are populated proportionally to their statistical weights. Also, each doublet level is split into the sublevels of the hyperfine structure because ^{10}B atoms have a nuclear spin distinct from zero ($I = 3$). Hence, the levels of the $2p\,^2P_{1/2}^0$ doublet split into two sublevels of the hyperfine structure and $2p\,^2P_{3/2}^0$ splits into four sublevels. Only one of the six mentioned sublevels can be used as the low level in the scheme of the two-photon excitation. In this work, we consider only

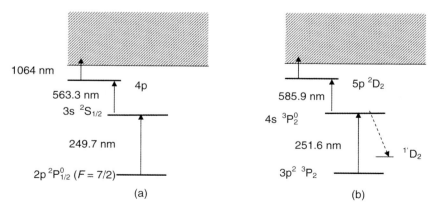

Fig. 5.4 Photoexcitation of (a) boron and (b) silica atoms.

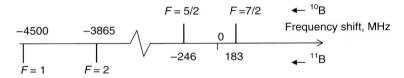

Fig. 5.5 Disposition of energy sublevels of the ground state of a boron atom.

sublevels of the lower level of the $2p\,^2P_{1/2}$ doublet with a simpler structure of hyperfine splitting.

The relative shift of the two sublevels for the two isotopes is schematically shown in Fig. 5.5. The corresponding data on the hyperfine structure of a resonance line for ^{11}B and ^{10}B are given in [187, 188]. But the values of isotopic shift presented in these works slightly differ. In the present work, we use the lower value. The selectivity rises at a greater energy difference between the hyperfine structure components of separated ^{10}B isotope and the components of ^{11}B isotope. One can see in Fig. 5.5 that the component with the total rotational momentum $F = 7/2$ has an advantage over the other component with $F = 5/2$. Also, the level with $F = 7/2$ has a greater statistical weight. It is this state that is suggested as the first low level in the scheme of excitation.

For the second state, the resonance level $3s\,^2S_{1/2}$ with a large absorption cross-section in photoexcitation and the lifetime of 4.1 ns is chosen [187]. In contrast, the third state should be long-living because in this case the losses to spontaneous emission are lower and the requirements for the power of ionizing radiation are not so rigid. For the third state, the level $4p\,^3P_{1/2}$ with the lifetime of 210 ns is suitable [187]. This choice, however, is not unique and strongly depends on the ionization efficiency.

Long-living Rydberg states can also be excited. In this case, the power of the photoionizing radiation may be noticeably reduced at the sacrifice of longer pulse duration.

Similar to the calculations for led, the main goal is to determine radiation parameters corresponding to the maximal productivity of the system. It is necessary to find the conditions, under which in the whole active medium the maximum possible number of high-purity ^{10}B and ^{28}Si isotope ions would be produced during the action of a single radiation pulse. The radiation energy on the first transition is limited by approximately 0.1 mJ. On other transitions this parameter may be an order of magnitude greater.

The case of an ideally collimated atomic beam was considered. Data on the lifetimes and oscillation strengths of levels were taken from [187].

The radiation pulses for the first and second transitions were assumed to be rectangular in shape with the duration of 10 ns. The characteristic ionization time for the third level was equal to 10 ns. The ionizing radiation pulse started just after the termination of the exciting pulse.

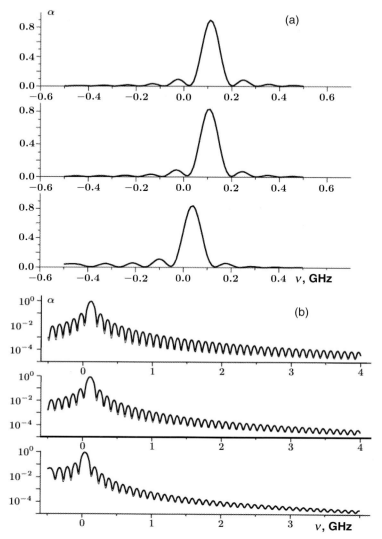

Fig. 5.6 Frequency dependence of the degree of ionization: (a) linear scale; (b) logarithmic scale. The upper, middle, and bottom curves correspond, respectively, to the energies at the first transition 30, 15, and 3 µJ cm^{-2}; the energies at the second transition are 30, 15, and 15 µJ cm^{-2}; the detunings are 2.19 GHz, 1.1 GHz, and 0.49 GHz.

The efficiency of the two-photon excitation depends on the detuning of the resonance frequency and, more strongly, on the deviation (the deviation of the frequency sum of the transitions from the frequency sum of laser radiation). The frequency dependence of the degree of ionization of boron isotopes is shown in

Fig. 5.6. Similar dependences are obtained for silica isotopes. It was assumed that the radiation frequency is fixed for the first transition and varies for the second transition. The abscissa is the frequency deviation. The oscillations on the frequency characteristic are specific for coherent interaction of radiation with atoms for both single- and two-photon excitation. By the logarithmic curve (see Fig. 5.6b) one can make the upper estimate of the ionization degree for the ^{11}B atom.

For a set of radiation energies there exists the optimal detuning of the resonance frequency and optimal deviation. The optimal deviation and the degree of ionization of boron isotopes are presented against detuning in Fig. 5.7. There is a maximum degree of ionization for ^{10}B atoms. Undesirable ionization of ^{11}B atoms sharply falls for detunings less than the optimal value. Hence, in the case of boron, for the detuning we may choose its optimal value by the degree of ionization. In this case, as one can see from Fig. 5.7, a high selectivity is provided.

Similar results are obtained for silica atoms. The silica spectrogram of excited even isotopes almost coincides with that of boron shown in Fig. 5.6. Since the isotopic shifts for silica are an order of magnitude less, the photoionization selectivity is much less than in the case of boron. This is illustrated by curves in Fig. 5.8. It was assumed that the radiation lines correspond to the transitions $3p^2\,^3P_2 \rightarrow 4s\,^3P_2^0 \rightarrow 5p\,^3D_2$ of ^{28}Si isotope.

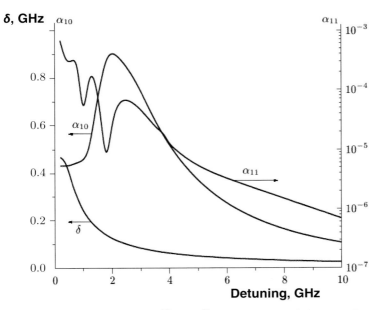

Fig. 5.7 Degree of ionization for ^{10}B and ^{11}B and the optimal deviation δ versus the detuning at the radiation energy of 30 µJ cm^{-2}.

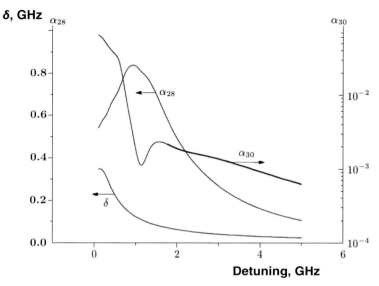

Fig. 5.8 Degree of ionization for ^{28}Si and ^{30}Si atoms and optimal deviation δ versus the detuning at the radiation energy of 15 µJ cm^{-2}.

Calculations show that

- the degree of ionization may reach 0.9;
- the half-height width of the spectral dependence of ionization at the optimal detuning and deviation weakly depends on the pulse energy and is approximately 100 MHz;
- the optimal deviation rises with the energy.

The most important result is that the width of the spectral dependence of the degree of ionization is ∼100 MHz. It is far less than the distance from the exciting line to the nearest line of ^{11}B isotope. This fact explains the preferable ionization of ^{10}B atoms.

A high degree of ionization of boron and silica atoms (about 0.9) can be obtained in a wide range of energies by appropriately choosing the detuning and the deviation. Calculations show that the energy of pulses 30 µJ cm^{-2} is acceptable. In this case, the detuning is about 2 GHz and the deviation is ∼0.1 GHz. The atomic excitation probabilities for ^{10}B and ^{28}Si are close to 0.9, for ^{11}B it is close to 0.000 03, and for ^{30}Si it is close to 0.02. Hence, the purification efficiency for boron may be very high, whereas for silica it is a few orders of magnitude less.

The system productivity is determined by the radiation power. According to calculations, under ideal conditions the degree of ionization may reach 0.9. Hence, approximately one quantum is needed in each cascade for producing single ion. From the pulse energy we can estimate the total number of ions in the whole volume. Assume, for example, that the power in the first transition, that is in the range of 250 nm, is 1 W at the pulse repetition frequency of 10 kHz. This corresponds to the origin of 1.2×10^{18} ions per second. In the case of the total ion extraction, we obtain the limiting productivity of 0.07 g h^{-1} for boron and 0.2 g h^{-1} for silica. Real values, however, are less. The radiation intensities would differ for different domains in the active volume due to longitudinal and transversal nonuniformity of the beam. Consequently, the degrees of ionization would also differ. Also, in the process of ion extraction, a part of them would be lost on the surfaces of the extractor elements used for protecting collector from atoms of undesirable isotopes. The efficiency of the two-photon excitation also falls due to the difference in the dipole momenta for excitation channels with different numbers of momentum projection. These factors reduce the limiting productivity at least twice. Hence, it is reasonable to assume the productivity at most 0.03 g h^{-1} and 0.1 g h^{-1} per 1 W radiation power in a CW operation mode.

5.4
Photochemical Separation of Zinc Isotopes by Means of the Two-Photon Excitation

5.4.1
Description of the Method

Similar to many other elements, zinc falls in the category of elements whose isotope selection in weight quantities by the selective laser photoionization is difficult. The main reasons are bad accommodation even on cold surfaces and the absence of nearby autoionization states. In [46, 47], the zinc isotope separation was experimentally performed by a different method, based on the two-photon excitation of the Zn atom to the state 6s $^3S_1$, however, by absorbing two oncoming photons ($\lambda_1 = 307.6$ nm and $\lambda_2 = 303.6$ nm) (see Fig. 4.14). The laser complex described in Chapter 2 was used for the excitation. Close energies of these photons in a two-photon process provide a low Doppler broadening down to 16 MHz, which is far less than the isotopic shifts. Then the state 6s $^3S_1$ decays into 4p $^3P^0_j$ in a series of spontaneous and induced transitions either direct or through the intermediate levels 5p $^3P^0_{0,1,2}$ and 5s $^3S_1$ (see Fig. 4.14). Intercombination transitions between systems of singlet and triplet levels of the Zn atom are very weak; hence, after the two-photon excitation occurs, all atoms are in the 4p $^3P^0_j$ state due to fast (the duration is shorter than 30 ns) processes. To prevent the direct isotopically nonselective excitation of Zn atoms in the single-photon process of pumping the first intercombination transition, the frequency of the laser radiation was shifted from the exact resonance of this transition by 4–5 half-widths of the Doppler profile ($\delta v_1 = (4-5)\Delta v_D$).

For selecting the desired isotope, the following exothermic photochemical reaction was used:

$$Zn^*(4p^3P_{0,1,2}) + CO_2 \rightarrow ZnO + CO + 1.3 \text{ eV} \tag{5.2}$$

The rate constant of reaction (5.2) was measured in a gas-circulating cell by the rate of CO production, which, within the measurement error (20 %) is independent of temperature $k^* = 2.5 \times 10^{-10}$ cm^3 s^{-1}. The rate constant of reaction involving Zn atoms in the ground state and CO_2 molecules is expressed by the formula

$$k \text{ [cm}^3/\text{s}^{-1}] = 7.9 \times 10^{-8} \exp(-10\,886/T \text{ [K]}) \tag{5.3}$$

At working temperatures ($T \approx 350$ °C this value is 4–5 orders of magnitude less than the rate constant of reaction with the excited atom. This provides a high selectivity of the process. At the CO_2 pressure of 0.1 Torr, the probability of the considered reaction is an order of magnitude greater than the decay probability for $(4p\,^3P_j^0)$-state, which is 10^5 s^{-1}. The quantum efficiency is greater than 50 %. This reaction is promising for the laser separation of Zn isotopes; nevertheless, it has significant drawbacks. The fact is that the two-photon resonance to the $(6s\,^3S_1)$-state is anomalously broadened (35 MHz Torr^{-1}) due to resonance processes that occur in impacts with CO_2 molecules (see Chapter 3). This value corresponds to the rate constant of broadening $k = 6.2 \times 10^{-9}$ cm^{3}s-1, which exceeds the rate constant of the chemical reaction. Despite the fact that finally the atoms are in the proper $(4p\,^3P_j^0)$-state, the results of the process are certain loss in selectivity and, at a greater degree, fall in the excitation efficiency due to the broadening of the absorption profile.

The product of chemical reaction (5.2) is the stable compound ZnO deposited on a collector. Produced CO molecules are evacuated from the separation zone.

The isotope separation was performed with the longitudinal or transversal circulation of gas mixture comprising CO_2 and Zn (see Fig. 4.3) at a relatively high pressure of CO_2 (~ 1 Torr). Since the concentration of Zn atoms is small ($\sim 10^{13}$ cm^{-3}), for enhancing the productivity in the real operation, a higher rate of pumping Zn through the working zone of a separation module is required. At an average power of 2 W, the number of reacted atoms N_a that is close to the number of absorbed photons (without allowing for the quantum efficiency and other processes) is $N_a = P_a/h\nu$, where P_a is the average radiation power absorbed and $h\nu$ is the quantum energy of the resonance transition. At $P_a = 2$ W, the number of reacted atoms per second is $N_a = 3.2 \times 10^{18}$, which corresponds to the circulation rate of 320 l s^{-1}. The pressure of gas reagent is 3–4 orders of magnitude greater than that of Zn vapor, which introduces difficulties in their circulation in a common flux. This is why the two-dimensional circulation was used in this work. The working mixture circulated along the beam at a velocity of ~ 2 l s^{-1} and Zn atoms circulated across the beam from the evaporator to a cold wall. In the intermediate zone of the Zn circulation, a collector was placed for accumulating the product in the form of isotopically changed zinc oxide. Further extraction of the isotopically enriched Zn is made by chemical methods.

A characteristic consequence of the efficient excitation of an atom to high states at high concentrations and large dimensions of the excitation zone is superradiation (generation of laser radiation without cavity), that is, the induced transfer to low levels. This effect limits the density of the excited atoms and, consequently, the productivity of laser isotope separation because of the broadening of absorption lines. It was established in experiments that this process is efficient under the condition

$$\sigma N^* L \geq 10 \tag{5.4}$$

where σ is the cross-section of the optical transition from the excited to the low state; N^* is the concentration of excited atoms; L is the characteristic dimension of the excitation zone. A mathematical model of this phenomenon will be given below.

Let us consider an example of the two-photon excitation of Zn atoms to the (6s $^3S_1$)-state. This state may efficiently decay into (5p $^3P_j^0$)- and (4p $^3P_j^0$)-states (see Fig. 4.14). The corresponding cross-sections of the transitions are 4×10^{-11} cm^2 and 3×10^{-13} cm^2. In this case, the efficient origin of superradiation occurs at the concentration $N^* \geq 5 \times 10^{10}$ cm^3 and at $L = 50$ cm.

Along with negative consequences, superradiation has some positive features. It is possible to automatically bind the frequency of laser radiation to the sum frequency of the two-photon excitation of one of the isotopes by the maximum of this parameter.

5.4.2
Polarization of Radiation

The photoexcitation scheme suggested (4s $^1P_0$ →4p $^3P_1^0$ →6s $^3S_1$) would never do for parallel polarization of the radiation. Such a transition is forbidden by selection rules. In plane-parallel light, quantum magnetic numbers are equal ($M = M'$); this is specific for certain channels of photoexcitation. The transitions with $M = M' = 0$ and $J = J'$ (where J is the angular momentum quantum number) are forbidden. The first transition in the suggested scheme selects the only channel with $M = 0$, that is, for all these transitions the equality $M = M' = 0$ should hold. The second cascade implies $J = J'$; hence, such a transition is forbidden. This selection rule can be surmounted by using light with circular polarization [63], which, however, is difficult to realize. It is very difficult to control the ellipticity of radiation and the quality of $\lambda/4$ plates used for obtaining circular polarization. The efficiency of absorbing radiation in the active zone strongly depended on the orientation of the plates.

Theoretical consideration shows that two plane polarizations crossed at an angle of 90° can be used instead of the circular one. This possibility can be explained qualitatively in the following way. Let the polarization planes of the radiations in the first and second transitions be perpendicular to each other. Consider two systems of coordinates with the quantization axes directed along the polarization directions in the first and second transitions, respectively. According to transformation rules for the ψ-function [189], in transferring from one coordinate system to another the state with $M = 0$ in the second level would transfer to the combined state with $M = 1$ and $M = -1$ in a new coordinate system (see Fig. 5.9). The plane-polarized radiation in the second transition excites these states to the levels with the same values of M. Hence, in using two waves with crossed plane polarizations it is possible to excite the third level.

One may think that five levels are involved in the excitation process (see Fig. 5.9). If this were the case, the number of equations for the density matrix elements would increase. Actually, this difficulty can be avoided by using the following formal trick.

It is known from the quantum mechanics [189] that a set of eigenfunctions for degenerated states can be substituted with an equivalent orthonormal basis composed of linear combinations of eigenfunctions. In quantization along the direction of light propagation we may use such a set of eigenfunctions, for which the radiation polarized along the x-axis would excite only one component of the basis in the second level. For this purpose, we may take the following system of functions:

$$\begin{aligned} \varphi_{-1} &= i/2(\psi_1 - \psi_{-1}) + 1/\sqrt{2}\psi_0 \\ \varphi_0 &= i/\sqrt{2}(\psi_1 + \psi_{-1}) \\ \varphi_1 &= i/2(\psi_1 - \psi_{-1}) - 1/\sqrt{2}\psi_0 \end{aligned} \quad (5.5)$$

The value of the quantum magnetic number is used as an index. It is easy to verify that system (5.5) really forms an orthonormal basis. In the old set, two components, namely, ψ_1 and ψ_{-1} were excited, whereas now only one component

$M = -1 \quad M = 0 \quad M = 1$

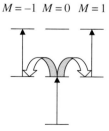

Fig. 5.9 Transitions with crossed polarizations. The shaped arrows denote the transformation of state in turning the system of coordinates.

φ_0 is excited. Indeed, the dipole momentum $\langle\varphi_i|d_x|\psi_g\rangle$, where ψ_g is the psi-function of the ground state, is distinct from zero for the only transition to the state φ_0:

$$\begin{aligned}\langle\varphi_0|d_x|\psi_g\rangle &= \langle i/\sqrt{2}(\psi_1+\psi_{-1})|(d_{-1}+d_{+1})/2|\psi_g\rangle \\ &= \langle i/\sqrt{2}\psi_1|d_{-1}/2|\psi_g\rangle + \langle i/\sqrt{2}\psi_{-1}|d_{+1}/2|\psi_g\rangle \\ &= (i/2\sqrt{2})(d_{-1}+d_{+1}) = i/\sqrt{2}d_{-1}\end{aligned} \qquad (5.6)$$

Here d_{-1} and d_{+1} are the dipole momenta for the left and right circular polarizations, respectively.

It can be shown for the remaining states that the dipole momentum is equal to zero. Similarly, in exciting the third level by the y-polarized radiation, the only transition $\varphi_0 \rightarrow \psi_0$ is distinct from zero; its dipole momentum is

$$\langle\psi_0|d_y|\varphi_0\rangle = -1/\sqrt{2}d_{-1} \qquad (5.7)$$

where ψ_0 is the state of the third level with $M=0$. For the corresponding transitions (the indices are omitted) we have

$$d_x = 1/\sqrt{3}d \qquad d_y = 1/\sqrt{3}d \qquad (5.8)$$

where the reduced dipole momentum is calculated by the known formula via the oscillator strength f (by absorption) [190]:

$$d^2 = (3he^2/2M_e)f(2J+1)/\nu \qquad (5.9)$$

Hence, the two-photon excitation can be calculated by the scheme $1\psi_0 \rightarrow 2\varphi_0 \rightarrow 3\psi_0$ without allowing for additional levels with different quantum magnetic numbers. The initial system of equations for the density matrix remains the same. Equations for radiation field intensities do not change as well, because for different polarizations they depend on d_x and d_y, respectively. It can easily be shown that the matrix elements of perturbations in the case of circular polarizations with equal radiation intensity exactly coincide with those for the case of plane polarizations. Hence, the linear polarizations are comparable with circular polarizations in the limiting efficiency and have an advantage of simple technical realization.

Two crossed plane polarizations can be obtained by means of a simple system of optical mirrors. Practical employment of plane crossed polarizations simplifies the optical scheme and results in a higher efficiency of isotope separation.

5.4.3
Mathematical Model of Cascade Superluminescence

Practical realization of laser isotope separation showed that in the two-photon excitation of zinc atoms the absorption of radiation nonlinearly depends on the atomic concentration. The reason for such a behavior is the origin of cascade superluminescence. Several lines were observed experimentally in the ranges 2.4 µm, 1.3 µm, and 0.48 µm. The power measured in the range 1.3 µm was about 100 mW.

Evolution of generation leads to an undesirable broadening of the absorption line while absorbing the pumping radiation. As a result, the efficiency of the two-photon excitation becomes several times lower. This effect forces operation at a relatively low concentration of atoms.

The result of the two-photon excitation is that dissipation channels for excitation energy open due to spontaneous and induced transitions. In Fig. 4.14, 12 possible transitions are shown. In some transitions, the population inversion may arise, whereas in other transitions it is not observed. For example, no superluminescence was observed in three lines of the transitions $6s\,^3S_1 \rightarrow 4p\,^3P^0_{1,2,3}$.

The transitions mentioned form a sufficiently complicated configuration. For formal calculation it is necessary to add 12 equations for the photon density and 7 equations for the inversion population (in the noncoherent approximation) to the system of equations for the density matrix. But in the considered case, the initial model inevitably becomes complicated and the calculations become time-consuming.

The noncoherent model was employed for calculating the power of lines and the population of levels. Since all atoms in the two-photon process with oncoming radiation beams are excited in a similar way, we may expect that the width of superluminescence lines would be close to the Doppler width. The coherence time of radiation cannot exceed the inverse spectral width and proves to be shorter than the pulse duration. Also, the coherence time of superluminescence should not exceed the time needed for the radiation to pass the active zone. Hence, radiative transitions can be characterized by the transition cross-sections similar to the incoherent case.

The kinetic of population can be considered in the saturated power approximation [191], which implies that the equality $N_2/g_2 = N_1/g_1$ holds, where indices 1 and 2 refer to the upper and lower levels, respectively. Nevertheless, in our case it is desirable to use the mild condition:

$$\frac{d}{dt}(N_2/g_2 - N_1/g_1) = 0 \tag{5.10}$$

that is, the populations of levels per unit statistical weight may differ by a constant value. Indeed, the saturated power approximation follows from the equation for photon density:

$$\frac{d}{dt}n_{\text{ph}} = P - n_{\text{ph}}/\tau \tag{5.11}$$

$$P = c\sigma n_{\text{ph}} g_2 (N_2/g_2 - N_1/g_1) \tag{5.12}$$

$$\sigma = \frac{1}{4\pi}(\ln 2/\pi)^{0.5} A\lambda^2/v_D \tag{5.13}$$

where P is the rate of photon origin due to induced transitions; τ is the photon lifetime in the active medium; σ is the cross-section of induced transition; A is the spontaneous emission probability; ν_D is the width of the Doppler profile. The derivative in (5.11) is estimated as n_{ph}/τ_p, where τ_p is the pulse duration, which can be neglected in comparison with n_{ph}/τ on the right-hand side of equation (5.11) because $\tau_p \gg \tau$ (τ does not exceed the time of radiation passage through the active zone). Then we have

$$P = \frac{n_{ph}}{\tau} \tag{5.14}$$

which entails the threshold condition:

$$\frac{N_2}{g_2} - \frac{N_1}{g_1} = \frac{1}{\tau c \sigma g_2} \tag{5.15}$$

The population equations may be written in the form

$$\frac{d}{dt} N_2 = G_2 - P \tag{5.16}$$

$$\frac{d}{dt} N_1 = G_1 + P \tag{5.17}$$

where $G_{1,2}$ is responsible for the rate of the level population change due to all processes involving the transition considered except for induced transitions. In view of (5.15), by adding (5.16) and (5.17) we find the equations for the populations without taking into account induced transitions. If we divide (5.16) by g_2 and then subtract (5.14), then using (5.15) we obtain

$$P = (G_2/g_2 - G_1/g_1) \frac{g_1 g_2}{g_1 + g_2} \tag{5.18}$$

The threshold condition (5.15) imposes limitations on some transitions. Two sets of transitions starting and ending at the same levels will be identified as conjugated cascades. From (5.15) it follows that the induced radiation is possible in the conjugate cascade, for which the parameter $\sum 1/(\tau c \sigma g_2)$ takes minimum (the sum is taken over all transitions of the cascade). Taking into account the fact that $A \sim (g_1/g_2)(f/\lambda^2)$, where f is the oscillator strength with respect to absorption, we obtain $\sigma g_2 \sim f g_1/\nu_D$. Then, the condition of minimum takes the form:

$$\sum (g_1 f/\nu_D)^{-1} = \min \tag{5.19}$$

Superluminescence on the transitions $6s\,^3S_1 \to 4p\,^3P^0_{1,2,3}$ is prohibited by condition (5.19). Lines corresponding to conjugated cascades were observed experimentally.

Condition (5.19) converts the graph of possible transitions to "woody" form, which substantially simplifies the calculation of radiation power in various transitions. Being sequentially applied, the procedure described above yields a set of linear algebraic equations for the radiation power. The matrix of the system can be inverted once and the radiation power can be automatically calculated for an arbitrary value of level population. The positivity of power should be verified in the process of calculation. If this verification fails, then the corresponding transition should be removed from the calculation.

The calculation becomes very simple in the case, where the kinetics of level populations in a cascade (except for the topmost level) is only determined by the processes of induced radiation. It can be shown that in the case of an unbranched cascade, the power of a particular transition from this cascade is calculated by the formula

$$P_{i+1} = P_i \frac{g_\Sigma}{g_\sigma + g_{i+1}} \tag{5.20}$$

where P_i is the rate of excitation due to induced transitions for the ith transition; g_i is the statistical weight of the upper level of the transition; g_Σ is the sum of the statistical weights of all levels down the cascade including the ground level of the transition. For the topmost level, the rate of excitation is determined by the intensity of the two-photon excitation. By the recurrent formula (5.20), the powers of all other transitions can be found. If generation occurs from one upper level to several lower levels, then the sum rate of exciting lower levels due to induced transitions is calculated by the formula similar to (5.20):

$$P_\Sigma = G \frac{g_\Sigma}{g_\Sigma + g_2} \tag{5.21}$$

where g_Σ is the sum statistical weight of all the lower levels; g_2 is the statistical weight of the upper level. The photon density is distributed proportionally to the statistical weights of lower levels.

The two cases considered lead to the general conclusion that in calculating the rate of populating a particular transition due to induced transitions, it suffices to calculate the statistical weight of all levels down the tree of the transition graph and to use formula (5.20) or (5.21).

Since we are interested in the sum frequency of induced transitions from the third level in the two-photon process, it suffices to calculate the rate of excitation due to the induced radiation by three transitions from the upper level and to use formula (5.18). In our case, we have $g_\Sigma = 21$ and $g_2 = 3$.

5.4.4
Calculation Results

With the aim of interpreting the experimental results theoretically and estimating the separation capabilities of the method, we performed a numerical simulation of the dynamics of level populations. The radiation lifetimes of working levels proved to be greater than the duration of laser radiation pulse (5–15 ns).

Hence, correct estimates necessitate the employment of the approximation of coherent interaction between radiation and matter. The Doppler distribution of atoms over velocities was taken into account.

According to the calculations performed, the degree of exciting zinc atoms to the long-living (4p ^{3}P$^0_{1,2,3}$)-state may be high (0.7) for oncoming radiation beams. This state is formed due to radiation transitions starting from the upper level. As was mentioned above, for every detuning there exists the optimal deviation. Atoms moving at different velocities have noticeably different detunings. The variation of the deviation in this case depends on the difference in the wavelengths for the first and second transitions. If the deviations of different atoms within the Doppler profile are close to the optimal value, then a high degree of excitation would be provided for all the atoms. The criterion of the efficiency of the employment of oncoming beams may be expressed in the form $d\delta_{opt}(\Delta + \nu - \nu_0)/d\nu \sim (\lambda_2 - \lambda_1)/\lambda$, where ν_0 is the central radiation frequency of the Doppler profile; Δ is the frequency detuning for the atom at rest (it is assumed that $|\lambda_2 - \lambda_1| \ll \lambda$ and the width of the Doppler profile does not exceed Δ). In the case of zinc, this condition is approximately satisfied in the range of operating energies. Since the optimal deviation falls with the detuning (see Figs. 5.7 and 5.8), it is desirable to provide negative wavelength difference in the first and second transitions, which is satisfied in the case of zinc atoms.

The calculated degree of isotope excitation at various concentrations of the latter versus the shift of the radiation frequency in the second transition is shown in Fig. 5.10. At a greater atomic concentration of isotope, the line broadens. This fact is explained by the superluminescence that occurs from the upper level to several lower levels, which results in line broadening. At a greater concentration, the absorption increases, and the population of upper level rises; hence, the superluminescence becomes more intense, and the line width increases. One can see that no superluminescence is developed in line wings; hence, it has no effect on the efficiency of exciting neighboring isotopes, from which the nearest one is separated by the frequency spacing of 660 MHz. Nevertheless, the separation selectivity falls because the degree of excitation at the center of line reduces.

In Fig. 5.11, the calculated degree of excitation, the absorption coefficient, and the degree of contamination by undesirable isotopes are shown against the concentration of the separated isotope. One can see that the absorption of radiation grows nonlinearly with the atomic concentration of isotope. Without superluminescence, the absorption increases linearly, whereas in the considered case the increase is noticeably retarded. The degree of contamination increases with the atomic concentration of isotope because of the lower degree of excitation of the separated isotope. The data presented satisfactorily agree with experimental results.

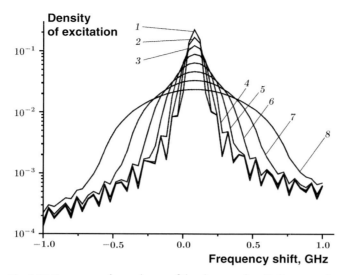

Fig. 5.10 Frequency dependence of the degree of excitation at various atomic concentrations of the separated isotope: (1) 2×10^{12} cm^{-3}; (2) 4×10^{12} cm^{-3}; (3) 8×10^{12} cm^{-3}; (4) 16×10^{12} cm^{-3}; (5) 32×10^{12} cm^{-3}; (6) 64×10^{12} cm^{-3}; (7) 128×10^{12} cm^{-3}; (8) 256×10^{12} cm^{-3}. The pulse energy is 250 µJ cm^{-2} and the detuning is 7 GHz.

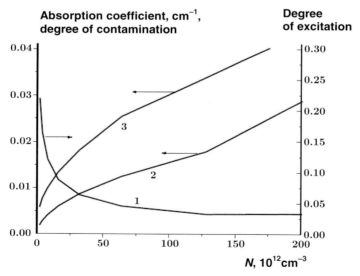

Fig. 5.11 Degree of excitation (1), absorption coefficient (2), and degree of contamination (3) versus the atomic concentration of isotope. The pulse energy is 250 µJ cm^{-2} and the detuning is 7 GHz.

5.4 Photochemical Separation of Zinc Isotopes by Means of the Two-Photon Excitation

Calculations show that 1% degree of contamination is obtained at the absorption coefficient at most 0.004 cm^{-1}. Experimental measurements of the absorption coefficient give close values.

5.4.5 Experimental Results

In Fig. 5.12, the experimental dependence of luminescence that arises after the two-photon absorption in the Zn transition 4s $^1S_0 \rightarrow$ 6s $^3S_1$ is shown against the detuning ν_2 in the second transition. At a low atomic concentration (see Fig. 5.12d) and moderate power of laser, the isotopic structure is completely resolved. The limiting spectral width of the separate isotope is 70 MHz and is determined by a finite width of the laser radiation line.

According to [182], the probability of the two-photon transition after averaging over atomic velocities in two oncoming radiation waves with frequencies ν_1 and ν_2 at small detuning from the intermediate level is determined by the expression

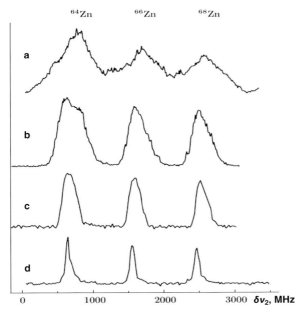

Fig. 5.12 Spectrum of luminescence at the wavelength of 1.3 μm versus $\delta\nu_2$ at $\delta\nu_1 = $ 5 GHz: (a) E_1, $E_2 = $ 550 μJ cm^{-2}, $n = $ 1.6 × 10^{13} cm^{-3}; (b) E_1, $E_2 = $ 400 μJ cm^{-2}, $n = $ 1.6 × 10^{13} cm^{-3}; (c) E_1, $E_2 = $ 150 μJ cm^{-2}, $n = $ 2.4 × 10^{14} cm^{-3}; (d) E_1, $E_2 = $ 150 μJ cm^{-2}, $n = $ 1.6 × 10^{13} cm^{-3}.

$$W = \frac{V_1^2 V_2^2}{4\Delta^2} \frac{\gamma}{(\nu_0 - \nu_1 - \nu_2 - \delta)^2 + (\gamma/2)^2} \qquad (5.22)$$

where, in the considered case, V_1 and V_2 are the Rabi frequencies for the single-photon transitions $4s\,^1S_0 \to 4p\,^3P_1^0$ and $4p\,^3P_1^0 \to 6s\,^3S_1$, respectively; γ is the width of the two-photon transition; the radiation frequency detuning Δ from the intermediate level is determined by the expression $\Delta = \nu_1 - \nu_{4s\,^1S_0 \to 4p\,^3P_1} = \nu_{4p\,^3P_1 \to 6s\,^3S_1} - \nu_2$; $\delta = (V_2^2 - V_1^2)/4\Delta$ is the field shift of the two-photon resonance frequency due to strong interaction between the radiation of frequency ν_2 and the optical transition. A calculated value of this shift is ≈ 200 MHz in the considered experimental parameters.

For the transition $4s\,^1S_0 \to 4p\,^3P_1^0$, the oscillator strength is more than two orders of magnitude less [188] than the oscillator strength of the transition $4p\,^3P_1^0 \to 6s\,^3S_1$. Consequently, we have $V_1 \ll V_2$. Hence, in calculating the shift of the absorption resonances for equal radiation powers for the case of two-photon absorption it is not necessary to take into account the transition from the ground state to an intermediate state. According to (5.22), the profile of the two-photon absorption line is symmetrical and its width is determined by various processes occurring in the separation zone. Nevertheless, one can see in Fig. 5.12 that the absorption lines possess definite asymmetry from the side of higher frequencies. This experimental fact is explained by fluctuations of pulsed power of laser radiation (up to 15 %). Indeed, if the profile of the two-photon line is detected during the action of single pulse, then it would be symmetrical as formula (5.22) predicts. Detection of the luminescence signal in a series of a large number of pulses results in an asymmetrical shape of observed resonances due to the field shift. Undoubtedly, this effect should be taken into account in efficient excitation of atoms for laser isotope separation because it leads to a loss in selectivity of chosen isotopes. The necessary condition for excluding such effects provides the experimental conditions under which the relationship $V_1 = V_2$ [192, 193] holds. In this case, the shift due to laser radiation is absent. Nevertheless, in the two-photon excitation of zinc atoms the fulfillment of the latter condition is not energy-justified, because to a single photon in the second transition would correspond a few dozens of photons in the first transition. In the two-photon excitation, only single photons participate in each transition. A considerable energy of the remaining photons in the first transition would be lost.

Formula (5.22) is proved by the fact that the estimates obtained with it correspond to the really observed degree of absorbing a monochromatic radiation in zinc vapors. Indeed, at the average power of propagating laser radiation ~ 1 W and the width of the absorption line 200 MHz, the probability of exciting by a single pulse is ~ 0.3. This value corresponds to the absorbed radiation power at the density of atoms $\sim 10^{13}$ cm^2 in the interaction zone.

At a higher concentration of Zn atoms (see Fig. 5.12c) or high density of radiation (see Fig. 5.12b), the lines of the two-photon resonance are strongly broadened or even overlapped (see Fig. 5.12a). The effect of absorption line broadening at an increasing atomic concentration of isotope or radiation intensity was experimentally observed and can be explained as follows. Under such growing, the part of absorbed radiation increases, which leads to a higher power of superluminescence and, consequently, higher frequency of induced transitions. Hence, the width of absorption lines increases. This effect leads to fall in selectivity. At a prescribed selectivity, the power density of laser radiation and the atomic concentration of separated isotope are bounded above.

The efficiency of the isotope separation is determined by the part of radiation absorbed in the process of two-photon pumping. It should be noted that in the two-photon absorption of two oncoming photons, all "useful" atoms participate in the process of absorbing the monochromatic radiation. The selectivity of the separation process is determined by the degree of overlapping of the lines of two-photon resonance. The latter parameter depends on the frequency detuning, radiation energy, and atomic concentration.

As was mentioned in Chapter 3, the atomic absorption lines may be considerably broadened in impacts with molecules of gas reagent. In Fig. 5.13, the experimental width of the absorption line of the two-photon transition $4s\,^1S \to 6s\,^3S_1$ for Zn atom is shown versus the pressure of such gas reagents as CO_2, CO, and NO. The corresponding broadenings are 35, 28, and 16 MHz Torr^{-1}. The cross-sections of this process are two orders of magnitude greater than the corresponding gas-kinetic values.

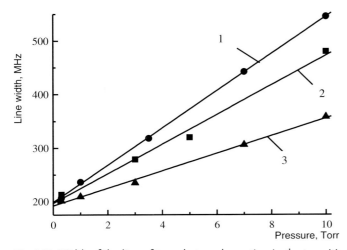

Fig. 5.13 Width of the line of two-photon absorption in the transition $4s\,^1S \to 6s\,^3S_1$ in zinc atom versus pressure of various gas reagents: (1) CO_2, (2) CO, and (3) NO.

This is explained by the fact that the energy of the transition $6s\,3S_1 \to 5p\,^3P_j$ (see Fig. 4.14) coincides with vibrational–rotational absorption bands. The result is that zinc atoms are found in the metastable $(4p\,^3P_j^0)$-state; however, the absorption line broadens. This limits the utmost concentration of gas-reagent atoms in the interaction zone.

The requirements for efficiency and selectivity contradict each other. Hence, each particular task necessitates a compromise between the specific productivity, efficiency, and selectivity of the separation process. For example, at the selectivity of 50–100, the density of radiation energy (E_1 and E_2) should not exceed 350 µJ cm^{-2} for $\Delta = 9$ GHz and the concentration of zinc atoms should not be greater than $N = 2 \times 10^{13}$ cm^{-3}. Under these conditions, the equivalent absorption coefficient for ^{66}Zn isotope is $k_\nu \sim (3–4) \times 10^{-3}$ cm^{-1}, which provides an efficient (up to 50%) employment of the pumping radiation at a moderate absorbing length (150–200 cm). The productivity of the complex is 0.6 g h^{-1}. In the considered case, it is possible to produce commercial product with the purity by target isotope at least 98% at the prime cost of approximately \$ 200 g^{-1}. The required separation selectivity is achieved by appropriately choosing the corresponding experimental parameters.

Hence, the two-photon excitation that is free of Doppler broadening used with the small detuning from the intermediate state combined with the photochemical reaction makes it possible to efficiently separate isotopes in the case where the resonance transitions of all separated isotopes lie within the Doppler profile of the absorption line.

Zinc isotopes are efficiently separated in gas centrifuges. The prime cost of isotopes obtained in this case is comparable to that obtained with either single- (see Chapter 4) or two-photon excitation in using laser methods. High real cost of isotope separation by laser methods is mainly determined by the low efficiency of CVL, short operational period of tubes and switches, and a small scale of installations. Calculations show that with large industrial complexes on the basis of Nd:YAG-lasers, the prime cost of zinc isotopes can be reduced by a factor of 3–5.

It is especially advantageous to employ the two-photon excitation in the case, where the isotope lines are masked by the Doppler profile and isotopes have no gaseous compositions (for example, Pd). High commercial price of Pd gives the possibility of obtaining a promising project profitability.

5.5
Zinc Isotope Separation by Evaporating Material from Chamber Walls

5.5.1
Problem Statement

We have considered the isotope separation schemes with gas circulation, in which the source of material vapor was outside the operation chamber. An alternative method for producing vapors inside chamber is to evaporate material that is deposited on the walls. This method is successfully employed in producing an active medium in copper-vapor lasers.

In [46], the heated tube was used, in which zinc pellets were placed before starting the operation. Gas reagent (CO_2) was pumped through the chamber for removing the reaction products (CO). Isotopically enriched ZnO was deposited on the chamber walls.

There was a slit in the tube for removing undesirable isotopes. Atoms, having passed through the slit, are precipitated on an external "cold" coaxial tube. Atoms of desired isotopes selected in the course of chemical reaction and precipitated on the walls of the internal tube as ZnO molecules had a lower probability to hit the slit.

It may be shown that at the pressure of gas reagent and, possibly, of inert buffer gas of at least a few tens of Torr, the diffuse mechanism forms the profile of atomic distribution of the separated isotope with respect to radius. This mechanism is responsible for a reduction of the atomic concentration at the center of tube. Variations of isotopic composition on the evaporating surface should also be taken into account. This variation occurs due to the participation of atoms of undesirable isotopes in numerous acts of evaporation from the surface and precipitation to it whereas the atoms of isotopes to be separated combined in molecules are firmly deposited on the surface. Hence, the near-surface layer, from which evaporation occurs is depleted of the desired isotope. A mathematical model was developed for revealing the role of these processes.

The slits on the internal cylinder break the axial symmetry. Generally speaking, the problem is two-dimensional. Solution of the two-dimensional nonstationary problem with strongly different characteristic times is rather difficult. However, it is possible to trace all principal features by a single-dimensional model and obtain numerical estimates.

We assume that the system operates in the mode of evaporating material from the inner surface of a chamber (see Fig. 5.14). We imply that the material only partially occupies the inner surface. Zinc atoms are partially excited during the action of radiation pulse and react completely with gas-reagent atoms in a time lapse between pulses, which means that the pressure of gas reagent is high enough.

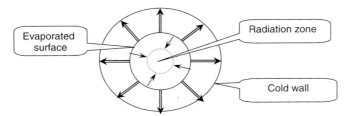

Fig. 5.14 Chamber with material evaporation from the surface of an internal cylinder. The thin arrows show the flux to the radiation zone and the thick arrows denote the flux to the external wall.

In the case of CO_2 this pressure, as was mentioned, is several tens of Torr. Since the excitation efficiencies differ considerably for different isotopes, the diffusion of two components is studied: atoms of the separated isotope and those of all the remaining isotopes as a whole.

It is assumed that the flux passing through the slit is uniformly distributed over the entire surface of the internal cylinder. Atoms passing outside the working cuvette are deposited on a cold wall.

The key moment in this model is the allowance made for the variation of isotopic composition on the surface of an evaporated material. The evaporation occurs from a thin surface layer with a thickness of several monatomic layers. The isotopic composition of the monatomic layer is taken into account by means of the substitution principle, which can be formulated as follows: atoms evaporated from the layer are substituted by the same number of atoms with the natural isotopic composition. This means that if the surface layer is enriched by a certain isotope, then, in the nonselective evaporation, the natural isotope composition would be restored. It was found that the thickness of the evaporated layer does not strongly influence the output characteristics if it varies within reasonable limits (see below).

Let us consider the boundary condition for the atomic concentration of the ith isotope on the metal surface. Let us denote the volume atomic concentration of the ith isotope near the surface by n_i and the atomic density of the near-surface evaporated layer (the number of atoms per unit area) by N_i. Let us also introduce the lifetime τ_c for atom on the surface until it is evaporated and the accommodation coefficient γ that is the probability of atom attachment to the surface in impacts with it. The value of τ_c is determined from the thermodynamic equilibrium condition:

$$N_0/\tau_c = \gamma n_0(T)v/4 \tag{5.23}$$

where index 0 refers to the sum concentration of atoms and v is the average atomic velocity. In particular, $n_0(T)$ is the atomic concentration for saturated vapors dependent on temperature T.

5.5 Zinc Isotope Separation by Evaporating Material from Chamber Walls

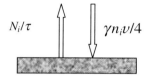

Fig. 5.15 Atomic flux from the surface and the backward flux of atoms from gas medium.

Taking into account the fluxes passing to and from the surface (see Fig. 5.15), we can write the equation for the surface atomic density N_i of the ith isotope in the following form:

$$dN_1/dt = -j_1 + \alpha_1 j_1 + \alpha_1 j_2 \tag{5.24}$$
$$dN_2/dt = -j_2 + \alpha_2 j_2 + \alpha_2 j_1 \tag{5.25}$$
$$j_1 = N_i/\tau_c - \gamma n_i v/4 \tag{5.26}$$

where j_i is the sum atomic flux of ith isotope passing from the surface, τ_c is the diffusion time related to the flux directed to the cold wall, and α_i is the part of the ith isotope in the natural zinc composition. The concentration N_i is taken near the surface. The last two terms in equations (5.24) and (5.25) are responsible for the change of atoms moved from the surface by those distributed according to the natural isotopic composition. In view of the relationship $\alpha_1 + \alpha_2 = 1$, equations (5.24) and (5.25) can be rewritten in the more compact form:

$$dN_1/dt = -\alpha_2 j_1 + \alpha_1 j_2 \tag{5.27}$$
$$dN_2/dt = -\alpha_1 j_2 + \alpha_2 j_1 \tag{5.28}$$

By adding (5.24) and (5.25) we obtain the equality

$$d(N_1 + N_2)/dt = dN_0/dt = 0$$

as expected, because the surface density of atoms remains constant.

On the other hand, the outgoing flux is equal to the sum of the diffusion fluxes inward chamber and outward it (to a cold wall):

$$\delta_s j_i = j_{int} + j_{ext} = D(\partial n_i/\partial r) + j_{ext} \tag{5.29}$$

where D is the diffusion coefficient; δ_s is the part of the area of the chamber surface occupied by evaporating material; j_{int} and j_{ext} are the densities of the diffusion fluxes inside and outside the internal tube, respectively. In the gap between the external and internal cylinders the distribution profiles are almost independent of time. The exact shape of the profile cannot be found; however, it can be estimated from solutions in the limiting cases. For example, if the gap between the cylinders is small as compared to the slit width, then the concentration distribution is close to the diffusion distribution between two coaxial cylinders:

$$j_{ext} = \delta_{sl} D n_i / (r \ln(R/r)) \tag{5.30}$$

where δ_{sl} is the part of the cuvette surface area related to the slits and R and r are the radii of the external and internal cylinders, respectively. We are more interested in the other limiting case, where the slit width is less than the separation between cylinders. Since there is really only one slit, the flux will be greater than that calculated by formula (5.30) and will approximately be equal to the flux

$$j_{ext} = 2\delta_{sl} D h_i / (h \ln(2d/h)) = D n_i / (\pi r \ln(2d/h)) \tag{5.31}$$

where h is the slit width and d is the separation between the external and internal cylinders. This approximate formula gives the value of the flux from the source with radius $h/2$ to the surface of the external cylinder. In numerical calculations, formulae (5.30) and (5.31) joint in such a way that at a small separation between cylinders formula (5.30) prevails and at a large separations formula (5.31) prevails. In analytical estimates we will employ formula (5.31), unless otherwise specified.

The diffusion equations for atomic components of composition are written in the standard form:

$$\frac{\partial n_i}{\partial t} = \frac{1}{r} D \frac{\partial}{\partial r}\left(r \frac{\partial n_i}{\partial r}\right) - \frac{n_i}{\tau_{conv}} \tag{5.32}$$

where τ_{conv} is the convection transport time of the particle determined as the ratio of the chamber volume to the volumetric rate of circulation.

Hence, equations (5.27) to (5.32) form the closed system describing the problem stated. It is worth recalling that we assume partial excitation of atoms during the radiation pulse and their total binding in the intermediate period in the result of chemical reaction. For solving this problem, a numerical method was elaborated, which was balanced in the number of particles.

5.5.2
Physical Analysis

In order to imagine the physical picture as a whole, let us compare the characteristic times of various processes participating in the separation of Zn isotopes. The fastest one is radiation excitation of atoms (~ 10 ns). In this time lapse, atoms actually do not change their positions and we assume this process instantaneous.

Let us derive the requirement for the concentration of gas-reagent molecules, at which most of the atoms participate in the chemical reaction during the lifetime of the excited state. From the characteristic rate of the chemical reaction (10^{-10} cm^{3}s^{-1}) and the lifetime of the excited state 10^{-5} s, we find $n_{th} > 10^{15}$ cm^{-3}. The interval between radiation pulses is an order of magnitude longer than the duration of the chemical reaction; hence, we may assume that all excited atoms react.

5.5 Zinc Isotope Separation by Evaporating Material from Chamber Walls

Table 5.1 Calculation parameters

Diameter of the external cylinder	8 cm
Diameter of the internal cylinder	4 cm
Length of the active zone	100 cm
Radius of the radiation zone	1 cm
CO_2 pressure	1 Torr
Concentration of Zn atoms	10^{13} cm^{-3}
Temperature	573 K
Diffusion coefficient	344 cm^{2}s^{-1}
Repetition frequency of pulses	10 kHz
Degree of excitation	0.1
Evaporation area to the total surface area ratio	0.2
Slit width	1 cm
Characteristic time of circulation	50 s

In what follows, physical estimates and conclusions will be illustrated by calculation results. At the beginning, let us calculate the system with the parameters presented in Table 5.1. Let us estimate the surface density of atoms by the formula (in the CGS system)

$$N_0 = (\rho N_A / M)^{2/3} \qquad (5.33)$$

where ρ is the specific mass, N_A is the Avogadro constant, and M is the atomic weight. By substituting the corresponding numerical values, we obtain the surface density $N_0 = 1.6 \times 10^{15}$ cm^{-2}. The accommodation coefficient was taken unity.

The next fast process is the establishing of the atomic concentration near the surface. The numerical calculation was performed, in which the atomic concentration in the active zone at the initial instant was zero. It proved that even before the second pulse, the atomic concentration near the surface was 0.92 relative to the concentration corresponding to the saturated vapor density. In a steady-state operation, the concentrations mentioned coincide within the accuracy of a fraction percent. The isotopic composition of vapor near the surface exactly equals the isotopic composition of the evaporated layer.

Let us estimate the characteristic transient period for the concentration of zinc atomic vapor corresponding to the saturated vapor pressure. Assume that at the initial instant the concentration of zinc atoms is zero. In a time lapse τ atoms will cover a distance that can be estimated as follows:

$$L = (4D\tau)^{1/2} \qquad (5.34)$$

By using condition (5.23) and the estimate $\partial n_i / \partial r = n_i / L$ we can rewrite (5.23) in the form

$$\frac{Dn_i}{L} = \frac{1}{4} \gamma v (n_i - n_{0i}) \qquad (5.35)$$

From (5.34) it follows that if the left-hand side of equation (5.35) is far less than $1/4\gamma v n_i$, then it may be taken zero; hence, the atomic concentration would correspond to the saturated vapor pressure for zinc. By using the known formula of the molecular gas theory $D = (1/3)\lambda v$, where λ is a mean free path, we can rewrite the latter condition in the form $\lambda \ll \gamma L$. Assuming, for definiteness, that λ is three times less than γL and allowing for (5.35) we obtain the estimate for the characteristic time:

$$\tau = 10\lambda/(\gamma v) \tag{5.36}$$

If we take the accommodation coefficient $\gamma = 1$, then, at a pressure of 1 Torr the characteristic time will be about $(2-5)\times 10^{-5}$ s, which agrees with the results of numerical calculations.

The next in the time growing scale is the characteristic transient time for the diffusion profile in the active zone. It is estimated by the formula

$$\tau_{\text{diff}} = 0.17 r^2/D \tag{5.37}$$

and under ordinary conditions equals $(1-30)\times 10^{-3}$ s. The time interval between pulses is too short for the profile corresponding to the regime without irradiation to develop. As a result, a dip arises in the central part of the atomic concentration distribution.

For a physical analysis it is useful to obtain the estimate of the diffuse lifetime for atom in the radiation zone of radius r_r:

$$\tau_r = \tau_{\text{diff}}(r_r/r)^2 \tag{5.38}$$

It follows from (5.38) that if the atom of the selected isotope hits the radiation zone, then it is subjected to the action of several pulses that occur in the time lapse τ_r. At the degree of excitation per pulse equal to 0.1, this means a sufficiently high probability for the atom to be excited.

The flux of atoms into the radiation zone determines the efficiency of excitation. Let us estimate it for the isotope to be separated.

We will characterize the efficiency of excitation by the effective lifetime of atoms with respect to excitation:

$$\tau_{\text{exc}} = -\frac{1}{f \ln(1-\delta_{\text{exc}})} \tag{5.39}$$

where f is the repetition frequency of pulses and δ_{exc} is the degree of excitation per single radiation pulse. It is easy to show that it is this characteristic time that is responsible for the reduction in the atomic concentration without additional replenishment of atoms. The diffuse distance characterizing the degree of atom penetration into the radiation zone is related to τ_{exc}:

$$\Lambda = (D\tau_{\text{exc}})^{1/2} \tag{5.40}$$

If $\Lambda \ll r_r$, then the flux through the boundary of the generation zone is estimated by the simple expression:

$$j_r = D n_r / \Lambda \tag{5.41}$$

where n_r is the atomic concentration for the isotope at the boundary of the radiation zone. This estimate is not perfect because it makes no allowance for the dependence of the flux on the radius of the radiation zone. A more accurate, however, cumbersome estimate can be made by formula (5.41) if Λ is substituted by the expression

$$\Lambda^* = \Lambda \frac{\exp(r_r/\Lambda) + \exp(-r_r/\Lambda)}{\exp(r_r/\Lambda) - \exp(-r_r/\Lambda)} \tag{5.42}$$

At $r_r/\Lambda \gg 1$ both the diffusion distances approach. Formula (5.42) illustrates the fact that in the contrary limiting case the flux tends to zero. While using (5.41) it is correct for a plane layer. In the axially-symmetrical case, the corresponding formula is in a complex manner expressed via Bessel functions of imaginary argument, yielding, however, no new physical meaning.

For determining the flux into the radiation zone we may bind the flux from an external zone with flux (5.41):

$$j_r = \frac{n_s - n_r}{r_r \ln(r/r_r)} \tag{5.43}$$

where n_s is the concentration near the surface.

From (5.41) and (5.42) one can easily find the parameter j_r on the surface of a cylinder with radius r_r and to recalculate it for the flux from the surface of the internal tube:

$$j_{int} = \frac{D n_s}{r \ln(r/r_r) + \Lambda^*} \tag{5.44}$$

The ratio of the atomic flux into the radiation zone to the total atomic flux is obtained from formulae (5.31) and (5.44):

$$\frac{j_{int}}{j_{ext} + j_{int}} = \left(1 + \frac{\ln(r/r_r) + \Lambda^*/r}{\pi \ln(2d/h)}\right)^{-1} \tag{5.45}$$

The latter relationship is responsible for the atomic proportion of the excited separated isotopes. The estimate by formulae (5.42) and (5.45) with the parameter taken from Table 5.1, yields $\Lambda^* = 0.61$ cm and $j_{int}/(j_s + j_{int}) = 0.81$. Numerical calculations yield $j_{int}/(j_s + j_{int}) = 0.69$. In view of the weak logarithmic dependence, the ratio of useful flux to the total flux mainly depends on the diffusion distance Λ^*. At shorter Λ^*, the part of useful flux increases. The easiest way to make Λ^* shorter is to reduce the diffusion coefficient by increasing pressure. But

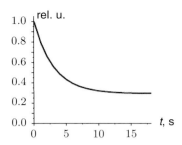

Fig. 5.16 Time evolution of the content of the separated isotope in the surface layer.

higher pressure makes characteristic diffusion time longer and reduces the total flux needed for high productivity of the system. Hence, a compromised choice of pressure is necessary.

The characteristic transient period for the isotopic composition in a thin evaporating layer of material is in the range of a few seconds. The temporal dependence of the part of atoms of the separated isotope ^{64}Zn evaporated from the layer relative to its natural composition is shown in Fig. 5.16.

One can see that the characteristic transient period is about 2.5 s (it corresponds to the half-difference between the initial and steady-state values). The steady-state values of isotope components are determined by the ratio of two atomic fluxes of separated isotopes. The first flux is directed to the center and is connected with the excitation of atoms, whereas the second flux is directed toward the cold wall due to the passage of atoms through cuvette slits. Let us estimate the steady-state isotopic composition in a near-surface layer. Assume that in this state the surface densities of the components do not change in time. Then, we may neglect the derivative in (5.27) and obtain

$$\alpha_2 j_1 = \alpha_1 j_2 \tag{5.46}$$

Boundary conditions (5.43) allow one to express the fluxes in (5.46) via the diffusion fluxes through the slit (5.23) and the flux directed to the center (5.44). Nevertheless, as was mentioned, the ratio of the component densities is equal to the ratio of the boundary atomic concentrations. In this case, the relationship between the components in the surface layer can be estimated as

$$\frac{j_1}{j_2} = \frac{N_1}{N_2} = \frac{n_1}{n_2} = \frac{\alpha_1}{\alpha_2}\left(\frac{1 + b_2/a}{1 + b_1/a}\right) \tag{5.47}$$

where

$$a = \frac{1}{\pi r \ln(2d/h)}, \quad b_i = \frac{1}{r \ln(r/r_r) + \Lambda^*} \tag{5.48}$$

5.5 Zinc Isotope Separation by Evaporating Material from Chamber Walls

From the viewpoint of the model adequacy it is important that the ratio of atomic concentrations near the surface is independent of such hardly identified parameters as the surface density of atoms or the lifetime of atoms on the surface. Numerical calculations completely prove this conclusion. Nevertheless, the transient period needed for the component density ratio to become steady noticeably depends on the parameters mentioned. But from the practical point of view, the transient period is not very important, because it is far shorter than the characteristic technological times. Without going into details, let us estimate the characteristic transient period for a steady state of isotopic composition under the condition that only selected isotopes are excited. The estimate yields

$$\tau_e = \frac{4\tau_{\text{diff}}}{\tau_s v \delta_s \gamma} \tag{5.49}$$

where the diffuse time is determined by formula (5.37), the lifetime for atoms on the surface τ_s is determined by (5.23), δ_s is the degree of filling the surface by material, v is the mean thermal velocity of atoms, and γ is the accommodation coefficient. From (5.49) it follows that the transient period for a steady state of isotopic composition is two or three orders of magnitude longer than the characteristic diffusion time, that is, under typical conditions it is in the range 1–10 s.

Let us now estimate the system productivity. From (5.44) and (5.47) in view of the boundary condition $n_1 + n_2 = n_0$ (where n_0 is the atomic concentration of saturated vapors) we may write the expression for the flux of separated isotopes (marked with index 1):

$$j_{\text{int}} = Dn_0 \frac{b_1}{1 + \frac{\alpha_2}{\alpha_1}\left(\frac{1+b_1/a}{1+b_2/a}\right)} \tag{5.50}$$

For the obviousness, formula (5.50) for the total flux of the separated isotope can be rewritten in the form

$$J_{\text{int}} = \frac{U}{R_1 + R_2 + R_3 + R_4} \tag{5.51}$$

where, in analogy with an electrical circuit, the "voltage" is

$$U = 2\pi D n_0 (1 + b_2/a) \tag{5.52}$$

and the "resistors", respectively, are

$$R_1 = \ln(r/r_r)/\alpha_1 \tag{5.53}$$
$$R_2 = (\Lambda_1^*/r)/\alpha_1 \tag{5.54}$$
$$R_3 = (\alpha_2/\alpha_1)\pi \ln(2d/h) \tag{5.55}$$
$$R_4 = (b_2/b_1)\pi \ln(2d/h) \tag{5.56}$$

The requirement of high selectivity means that the atomic flux of isotopes of the second kind through the slit should be more intense than the flux into the radiation zone. Hence, the necessary selectivity condition is the inequality $a \gg b_2$, which entails

$$\pi \ln(2d/h) \ll r \ln(r/r_r) + \Lambda_2^* \tag{5.57}$$

Then "voltage" (5.52) depends only on the diffusion coefficient and the saturated vapor concentration:

$$U = 2\pi D n_0 \tag{5.58}$$

Hence, the productivity would be determined by the values of four "resistors" connected in series.

5.5.3
Calculation Results and Their Analysis

The results of numerical and analytical calculations of the model variant with the parameters taken from Table 5.1 are presented in Table 5.2. The results may be extended to different values of n_0 because the degree of excitation is independent of n_0. All the fluxes proportionally increase with the rise of saturated vapor density. We may neglect the weak temperature dependence of the diffusion coefficient in the range of working temperatures.

The radial distribution of the atomic concentration of the separated isotope is shown in Fig. 5.17. One can see a strong dip in the profile of the atomic concentration distribution in the radiation zone (its radius is 1 cm).

The rate of producing molecules versus pressure is shown in Fig. 5.18. One can see a good agreement between the numerical simulation and the analytical calculation by formula (5.50). A difference is observed at a low pressure only. In the latter case, the atom passing to the radiation zone would not necessarily be

Table 5.2 Comparison of the results of numerical simulation and analytical calculation.

Parameter	Numerical calculation	Analytical calculation
Near-surface atomic concentration of separated isotope	0.082×10^{13} cm^{-3}	0.066×10^{13} cm^{-3}
Flux of the atoms of separated isotope into the radiation zone	0.91×10^{17} s^{-1}	0.93×10^{17} s^{-1}
Flux of the atoms of separated isotope passing though the slit	0.27×10^{17} s^{-1}	0.21×10^{17} s^{-1}
Flux of the atoms of remaining isotopes passing through the slit	3.02×10^{17} s^{-1}	2.92×10^{17} s^{-1}

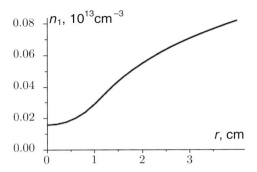

Fig. 5.17 Radial atomic distribution of the separated isotope.

excited. The surface of the radiation zone cannot be assumed ideally absorbing the atoms of separated isotope. At a low pressure, significant is "resistor" R_2 (see (5.54)) proportional to the diffusion distance Λ^* (see (5.42)), which becomes comparable to "resistor" R_1 of the coaxial system (see (5.53)). At a pressure of 1 Torr and higher, the parameter Λ^* may be neglected in comparison with $r \ln(r/r_r)$ (at the pressure of 1 Torr we have $\Lambda^* = 0.6$ cm). In this case, the useful atomic flux is proportional to the diffusion coefficient. Hence, at a high pressure we obtain a strongly inverse proportionality to the pressure. At a low pressure, the flux is

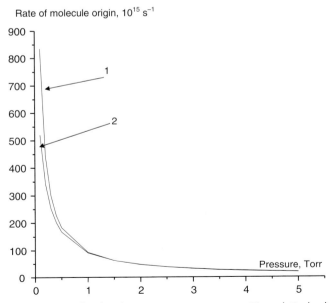

Fig. 5.18 Rate of molecule origin versus pressure: (1) analytical calculation; (2) numerical calculation.

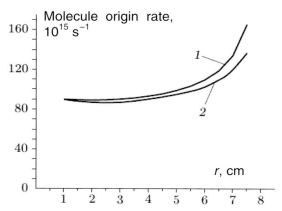

Fig. 5.19 Rate of molecule origin versus radius of internal cylinder: (1) analytical calculation; (2) numerical simulation.

slightly lower then it follows from the inverse dependence. Really, at the pressure of 0.1 Torr, the lifetime of the excited atom with respect to the chemical reaction becomes comparable with the duration of the interval between pulses, which results in a lower productivity. At the reaction rate $\sim 10^{-10}$ cm^3 s^{-1}, the optimal pressure is expected in the range 0.2–0.4 Torr. Hence, a considerable rise in productivity is observed at lower pressures. Nevertheless, the rate of molecule origin 3×10^{17} s^{-1}, almost maximal at a given geometry, is still low as compared to the flux of radiation quanta 1.6×10^{18} s^{-1} (at the power of 1 W).

The rate of molecule origin versus the radius of internal tube is shown in Fig. 5.19. One can see that this dependence is rather weak. Only if the tube radius is close to the radius of the outer cylinder, the rate of molecule origin rises. Such a behavior is explained by formula (5.51). From this formula it also follows that at a greater radius, "resistor" R_1 between the cuvette surface and the radiation zone increases, whereas R_2 and R_3 reduce. If the radius of cuvette becomes close to that of chamber, then "resistor" R_3 between the cuvette and the chamber surfaces reduces most rapidly. The result is that the rate of molecule origin rises.

The dependence of the molecule production rate versus the radius of the outer cylinder is shown in Fig. 5.20. Similar to Fig. 5.19, if the radii of the outer and inner cylinders are close, then the molecule production rate increases.

In Fig. 5.21, the rate of molecule origin versus the radius of radiation zone is shown. One can see an abrupt reduction in the range of radius lengths less than 1 cm. This is explained by the fact that the ratio of the probability for atom to hit the radiation zone to that of hitting the slit falls at a lower radius, because the width of the slit remains constant.

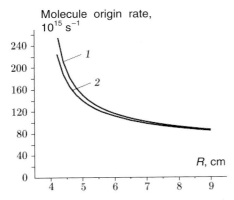

Fig. 5.20 Rate of molecule origin versus radius of external cylinder: (1) analytical calculation; (2) numerical simulation.

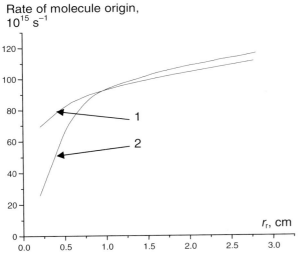

Fig. 5.21 Rate of molecule origin versus radius r_r: (1) analytical calculation; (2) numerical simulation.

5.5.4
Influence of Diffusion Processes on the Selectivity of Isotope Separation

As was shown, diffuse processes result in that under typical conditions the concentration of the selected isotope in the radiation zone is more than an order of magnitude less than the concentrations of other isotopes. The resulting selectivity of excitation related to the radiation processes is markedly reduced due to diffusion processes and the variation of isotopic composition in the surface layer. In spite of the previous calculations, we will assume that one more, useless isotope

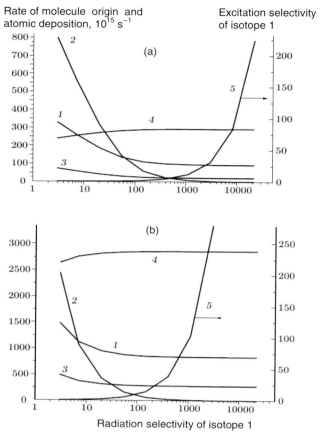

Fig. 5.22 The total fluxes for various isotopes at a pressure of (a) 1 Torr and (b) 0.1 Torr: (1, 2) the rates of producing molecules of first and second isotopes, respectively; (3, 4) the fluxes to the wall of the external cylinder for first and second isotopes, respectively; (5) the final selectivity of the process.

is also excited. At the parameters of the example variant (see Table 5.1) we will vary the radiation selectivity of excitation, that is, the selectivity without taking into account diffusion processes. As was shown in the previous investigations, the radiation selectivity for ^{64}Zn at the degree of atomic excitation of the selected isotope about 0.1 and at the absorption coefficient ~ 0.004 cm^{-1} may be greater than 100. At a lower absorption coefficient the radiation selectivity may become greater. Nevertheless, the final selectivity of the separation process is several times less.

In Fig. 5.22a, the total fluxes are shown for different isotopes from the whole cuvette of length 100 cm to the radiation zone and to the external space. The first flux corresponds to the rate of molecule origin and the second one relates to the rate of material deposition on a cold wall of the external cylinder. The behavior of these curves is well described by the estimates given in Section 5.5.2. It is important that the final selectivity strongly falls relative to the radiation selectivity. At a low pressure the situation is better. In Fig. 5.22b, similar data are presented, however, for the pressure of 0.1 Torr (lower pressure is not sensible because of the low rate of the chemical reaction). As one can see, the selectivity is noticeably higher in this case. The pressure dependence of the final selectivity is shown in Fig. 5.23 at the radiation selectivity of 100. Hence, obtaining the desired selectivity at a reasonable choice of the parameters is a problem. It can only be obtained by a repeated processing of material.

Qualitatively, the behavior of the curves can be described by the formula for the flux ratio (selectivity):

$$\frac{J_{s1}}{J_{s2}} = \frac{\alpha_1}{\alpha_2} \frac{\sqrt{D/(f\delta_{exc2})}}{\pi r \ln(2d/h)} \tag{5.59}$$

which follows from (5.40) and (5.53), the assumption $\delta_{exc1} \gg \delta_{exc2}$, and other less important assumptions. From formula (5.59) one can see that the reduction of the degree of excitation of the useless isotope by a factor of 25 results in only a 5-fold increase of the selectivity.

Hence, a drawback of the method considered in this chapter is a long time needed for renewal of isotopic composition in the active zone. This time is longer than the lifetime of atoms of the separated isotope with respect to the chemical reaction. The resulting isotope composition in the working volume and on the surface of the internal tube is depleted of the selected isotope. As a consequence, the productivity and the purity of the selected isotope fall. In spite of this drawback, this method is rather attractive due to the construction simplicity of the separation unit and may be competitive as compared to other methods in producing small quantities of product.

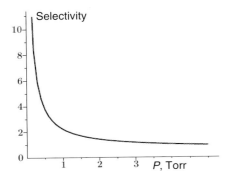

Fig. 5.23 The selectivity of isotope separation versus pressure.

6
Prospects for Industrial Isotope Production by Methods of Laser Isotope Separation

Employment of isotopes revolutionized many investigation methods in science and technology, opened new fields of knowledge and production. Wide employment of isotopes resulted in elaboration of numerous methods of isotope separation [1].

Employment of stable and radioactive isotopes started just after their discovery and presently covers almost all directions of human activity. Undoubtedly, the development of separation methods including LIS was actively stimulated by needs of nuclear power engineering, arm production, and medicine. In addition, employment of isotopes in other fields has reached the level exceeding billion dollars yearly. Some examples of "civilian" employment of isotopes are given in Table 6.1.

Employment of isotopes in nuclear power engineering and medicine is most profound. After nuclear power engineering, medicine expenditure on employment of stable and radioactive isotopes is more than a half in money terms. An obvious limiting factor of their extended employment is high cost and the difficulties of converting "traditional" isotope production plants from one product to another. For example, such a conversion in centrifugal or isotope exchange technology may require several months.

Laser methods of isotope separation seem to be most flexible and cover most of the elements of the periodic table. Due to their potential advantages (universality, low energy expenditure, short start-up period and so forth) [78], they prove to be relatively "cheap", "simple", and, consequently, promising for industrial isotope production.

Nevertheless, it should be taken into account that laser methods are introduced in highly developed fields of science and technique. A high competition in the isotope separation industry is partially caused by a considerable reduction in the part of the military procurement and highly developed alternative technologies for producing isotopically changed materials. Hence, it is reasonable to widely apply LIS for those elements and in those fields, for which the advantages of laser methods are obvious (for example, there are no alternative methods or it is more economically profitable as compared to other industrial methods).

Laser Isotope Separation in Atomic Vapor. P. A. Bokhan, V. V. Buchanov, N. V. Fateev,
M. M. Kalugin, M. A. Kazaryan, A. M. Prokhorov, D. E. Zakrevskiĭ
Copyright © 2006 WILEY-VCH Verlag GmbH & Co. KGaA, Weinheim
ISBN: 3-527-40621-2

Table 6.1 Fields of application and employment of isotopes

Field of application	Employment
Biology, biochemistry, biotechnology, physiology	Study of metabolism processes, structure and functions of biological molecules; biological tracers; study of mutations in living organisms
Cosmology	Study of origin and evolution of the universe, star systems, and planets
Medicine	Nuclear diagnostics of diseases; study of functions and functioning of organs; positron, γ- and NMR-tomography; radionuclide therapy of certain organs; study of functioning of medical products
Earth sciences: geology geochemistry, geophysics, hydrology; sea sciences	Determination of rock ages, radiometric field analysis; activation methods of search and storage of minerals; study of accumulation and migration of groundwater, dynamics of lakes and storage pools; control of river flow and ocean streams
Food	Study of food properties aimed at preventing cancer and cardiovascular diseases, dyspepsia, and metabolic diseases; study of metabolism energetics including certain organs
Industrial production and control of technological processes	Material science; determination of the velocity and expenditure of materials; determination of the thickness and structure of materials and goods; crack detection; study of mixing, burning, diffusion, and phase transformations; computer production
Agriculture	Growing high-yield plants capable of withstanding weather variations; livestock breeding; longer harvest preservation; study of photosynthesis, assimilability of fertilizers and microelements
Physicochemical sciences	Nuclear physical experiments; study of fundamental laws of the universe; activation analysis; study of chemical transformations and processes
Ecology and environment investigations	Study of global processes of the environment pollution, concentration of detrimental substances in food chains of biological organisms, greenhouse effect and mechanisms of ozone holes origin
Power engineering	Fuel for nuclear and thermonuclear power plants; construction and auxiliary materials for nuclear power plants

Isotope market is specific in that determining the demands for already produced and demonstrating the interest to new isotopes are often problematic. On the one hand, real interest should be separated from speculative interest that is caused by very high price of certain isotopes. On the other hand, interest shown to certain isotopes may reveal "know how" in high-end technologies, which is an industrial and commercial secret.

6.1
Microelectronics and Optoelectronics

From the above reasoning, authors mainly focus on separate reliably confirmed applications, which may form a basis for economically profitable production. Below, we describe some most promising applications of laser isotope separation in atomic vapors.

6.1
Microelectronics and Optoelectronics

Microelectronics traditionally imposes strict requirements on the purity of materials employed. Nevertheless, the interest in isotopically pure materials increased in the last decade only. It seems that three materials important in modern microelectronics are desirable in an isotopically changed state. These are silica, boron, and led.

Isotopically changed Si with the content of ^{30}Si greater than 70 % (in comparison with 3.1 % in natural silica) is needed for uniform doping with phosphorus by irradiating massive specimen by neutrons. In the result of the reaction ^{30}Si + n $\rightarrow$ ^{31}Si unstable isotope ^{31}Si arises, which, by β-decay transforms to phosphorus: ^{30}Si $\rightarrow$ ^{31}P + β. If silica specimen is uniformly irradiated, then a uniformly doped material with improved operational characteristics is obtained. Isotopic dopings are considered in Appendix D in more detail.

Highly enriched silica isotopes, in particular, ^{28}Si are used for obtaining materials with an enhanced thermal conductivity and in the developments of quantum computer [194–198]. Silica needed for these purposes should be isotopically pure up to 99.99 %.

Boron monoisotopes (usually ^{11}B) are used in the ion implantation for obtaining abrupt (p–n)-junctions. Free length of various boron isotopes is different, which results in blurring of the (p–n)-junction boundary. Employment of Boron in microelectronics is considered in Appendix E in more detail.

Similar to ^{11}B, both ^{28}Si and ^{30}Si can be obtained in large quantities by traditional methods of enrichment. Since ^{10}B is widely used in nuclear power engineering, industrial methods for producing it were developed [199]. These methods can be easily adopted to large-scale separation of ^{11}B with an acceptable prime cost. In recent years, the production of ^{28}Si with a high isotopic purity (from 99.9 % to 99.99 %) by means of gas centrifuges was intensively developed [200, 201]. Laser methods are advantageous, in particular, in producing highly pure ^{11}B and ^{28}Si at the final stage of traditional methods, because in this case the purification occurs with a moderate amount of laser energy (see Chapter 5). One more promising field of application is *online* isotope producing, where obtaining isotopically changed material is a constituent of the united industrial cycle.

Low radioactive led is used for producing microprocessors with the operation frequency of higher than 500 MHz by "flip-chip" technology. Chips are saturated with led, which is deposited in the form of semispherical bamps over the whole surface of the silica plate. The elements of a microprocessor, in particular memory elements, are irradiated by α-particles that arise in the decay of 210Pb according to the chain ^{210}Pb $\to$ ^{210}Bi + β $\to$ ^{210}Po + β $\to$ ^{206}Pb + α. At small dimensions of the element, which is a characteristic for high-performance microprocessors, penetration of α-particles disturbs the nonconductive state [202] and a spontaneous information error occurs. The higher the packing density, the more strict demands are imposed on the purity of led (with respect to ^{210}Pb) and other construction materials used in a microprocessor [203]. The up-to-date level of the requirements for such materials is $\sim 10^{-3}$ α cm^{-2} h^{-1}; the sum radioactivity does not exceed the equivalent space background ($\sim 2 \times 10^{-3}$ α cm^{-2} h^{-1}).

Low-radioactive led was obtained by laser methods [98, 99] and its production was realized on a semiindustrial scale [204]. The production cycle is divided into two stages. At the first stage, led ore is selected with a low content of natural radioactivity and PbS mineral (galenite) is separated. Then led melting is performed by the technologies that exclude additional radioactive contamination of the material followed by vacuum distillation or electrochemical purification. Such material is supplied to world market by corporations "Mitshubishi Material" with the declared radioactivity at most 0.05 α cm^{-2} h^{-1}, "Jonson Mathey Electronics" with radioactivity of 0.02 α cm^{-2} h^{-1}, and company "Pure Technologies" (with the production plant in Izhevsk, Russia) with led radioactivity of 0.01 α cm^{-2} h^{-1}.

Starting from the level of 0.05 α cm^{-2} h^{-1}, laser methods of led purification compete with the methods of separating low-radioactive led from natural ore, and at the level of 0.01 α cm^{-2} h^{-1} there are no alternative methods for producing led on an industrial scale. Such led (with the radioactivity less than 0.005 α cm^{-2} h^{-1}) is produced by company "Coherent Technologies" (Novosibirsk, Russia).

The price of the low-radioactive led is, depending on quality, $ 100–$ 500 kg^{-1}. By the early millennium, about 10^9 microprocessors were produced in the world by "flip–chip" technology. It is expected that their annual production will be $\sim 10^{10}$ microprocessors. The number of bamps in a single chip is of the order of 10^4 at the dimensions 50–100 μm. This corresponds to the technological expenditure of led $\sim$30 mg per chip or 300 t yearly. Even at the price of low-radioactive led $\sim$ $ 500 kg^{-1}, it will be $ 0.015 per chip, which is a negligible part from the total price of the microprocessor. The total cost of led would be $\sim$ $ 100 million per year, which is acceptable from the viewpoint of creating large-scale production. Nevertheless, the demand of preserving the environment seems to be a substantial limiting factor of the increasing production of the low-radioactive led [205].

6.2 Nuclear Fuel Cycle

In recent years, an interest was shown to isotopically changed led as a promising coolant in fast-neutron nuclear power plants. It is assumed that the heavy metal coolant and the proton-to-neutron converter will be used (led or led–bismuth alloy). Unfortunately, the long-term irradiation of led results in producing extremely dangerous α- and γ-radionuclides, including ^{210}Po and ^{207}Bi. It was shown that a coolant enriched by isotope ^{206}Pb (95 %) produces far less (by a factor of 10^3–10^4) polonium and bismuth radionuclides than natural led [206, 207]. Hence, it is desirable to study new possibilities for separating led isotopes (see Fig. 6.1) aimed at obtaining a considerable amount of ^{206}Pb (several hundreds ton) at an acceptable prime cost. The laser technology is suggested for producing large amounts of the isotope, which is based on "burning out" of selectively excited led atomic states due to photochemical selective reactions (see Chapter 5). Led atom has long-living metastable states, which is a promising factor in solving the separation problem. Preliminary economical estimates show that the prime cost of obtaining ^{206}Pb by laser photochemical methods would not be greater than $ 0.02 g^{-1} (for a comparison, in centrifugal separation the prime cost is $\sim$ $ 20 g^{-1}).

Fig. 6.1 Chamber for separating lead isotopes.

Table 6.2 Employment of isotopes in medicine

Isotope	Field of application (treatment, diagnostics, investigations)
^{225}Ac	Radiation therapy
^{211}At	Radiation therapy
^{10}B	Food tagging for determining the role of boron in metabolism
^{211}Bi, ^{213}Bi	Radiation therapy
^{77}Br	Radionuclide indicator
^{11}C	Positron tomography
^{13}C	Study of reactions in organic chemistry; isotopic breath tests; study of the molecular structure of large biological molecules, etc.
^{42}Ca, ^{44}Ca, ^{46}Ca, ^{48}Ca	Study of calcium metabolism; study of the role of addible calcium in pregnancy, juvenile growth and evolution; treatment for bone diseases
^{112}Cd	Production of ^{111}In for γ-diagnostics and therapy by open sources
^{35}Cl, ^{37}Cl	Toxicity investigations of external pollutants
^{50}Cr	Production of ^{51}Cr for γ-diagnostics
^{51}Cr	Gamma tomography
^{53}Cr, ^{54}Cr	Noninvasive study of chromium metabolism and the needs of human organism in chromium
^{63}Cu, ^{65}Cu	Noninvasive study of copper metabolism and the needs of human organism in copper; study of congenital disorders; study of tissue strength (including myocardium)
D(H)	Vitamin investigations; study of the mechanisms of chemical reactions
^{18}F	Positron tomography
^{52}Fe	Indication of blood circulation in human body
^{54}Fe, ^{57}Fe, ^{58}Fe	Study of metabolism; determination of energy consumption and control of genetic mechanisms; investigation and therapy of anemia; study of the conditions for iron assimilation
^{67}Ga	Gamma tomography
^{68}Ga	Positron tomography
^{76}Ge	Production of ^{77}As
^{3}He	Study of living organisms by NMR
^{166}Ho	Radiation therapy
^{123}I	Gamma tomography; radionuclide indicator
^{131}I	Gamma tomography
^{111}In	Gamma tomography
^{78}Kr, ^{80}Kr, ^{82}Kr, ^{84}Kr, ^{86}Kr	Diagnostics of lungs diseases
^{81}Kr	Gamma tomography
^{6}Li	Kidney physiology investigations; determination of membrane permeability; therapy of mental diseases
^{25}Mg, ^{26}Mg	Noninvasive study of human food requirements; study of metabolism and food assimilation; study of the evolution of cardiovascular diseases
^{52}Mn	Indicator of biological processes in organisms

Table 6.2 (continued)

Isotope	Field of application (treatment, diagnostics, investigations)
^{94}Mo, ^{96}Mo, ^{97}Mo, ^{100}Mo	External tagging of food for determining the human needs in food
^{99}Mo	Parent element for producing ^{99}Tc
^{100}Mo (purity 99.99 %)	Cyclotron target for producing ^{99}Tc
^{13}N	Positron tomography
^{15}N	Large-scale employment for investigating nitrogen circulation and assimilability in agriculture; synthesis; study of catabolism and protein circulation in organism; study of metabolism in tissues
^{57}Ni	Indicator of biological processes in human organism
^{14}O, ^{15}O	Positron tomography
^{17}O	Investigations in structural biology; study of cataract
^{18}O	Study of energy consumption by living organisms (from plants to human, comparative and relative); study of metabolism
^{32}P	Radiation therapy
^{203}Pb	Gamma tomography
^{103}Pd	Gamma tomography
^{81}Rb	Parent element for producing ^{81}Kr
^{82}Rb	Positron tomography
^{85}Rb, ^{87}Rb	Production of radioisotope ^{88}Sr; tracers of sodium metabolism; study of mental diseases
^{186}Re, ^{188}Re	Radiation therapy
^{33}S, ^{34}S	Study of human genome
^{73}Se	Radionuclide indicator in biological processes with selenium
^{74}Se, ^{76}Se, ^{77}Se, ^{78}Se, ^{80}Se, ^{82}Se	Study of the role of selenium in biological processes in human organism
^{153}Sm	Radiation therapy
^{113}Sn	Radioactive parent element for obtaining ^{113}In
^{117}Sn	Radiation therapy
^{82}Sr	Production of ^{82}Rb
^{89}Sr	Radiation therapy
^{90}Sr	Radioactive parent element for producing ^{90}Y
^{99}Tc	Gamma tomography
^{122}Te	Parent isotope for producing ^{131}I
^{123}Te, ^{124}Te	Parent isotope for producing ^{123}I
^{201}Tl	Gamma tomography
^{51}V	Study of diabetes, assimilability of biological products, and metabolism
^{188}W	Radioactive parent element for producing ^{188}Re
^{124}Xe	Parent isotope for producing ^{131}I
^{129}Xe	Magnetic resonance tomography
^{90}Y	Radiation therapy
^{64}Zn, ^{67}Zn, ^{68}Zn, ^{70}Zn	Noninvasive determination of zinc demands in organism; study of metabolic imbalance; study of liver diseases; study of assimilability and food demands

An interest in boron isotopes also increased. In particular, ^{10}B is used as a material for rods that control the rate of nuclear processes in slow-neutron reactors due to a large cross-section of neutron capture in a wide range of energies. Boron steels doped with highly concentrated ^{10}B are used in producing storages and transportation containers for nuclear fuel waist and in solving the protection problems of transporting nuclear fuel. Some additional information on employment of boron is given in Appendix F.

For the liquid control of nuclear reactivity it seems promising to use solutions of gadolinium nitrate with ^{157}Gd, because it possesses high cross-section of neutron absorption and high solubility in water coolant.

Isotopically changed zinc is being used to an increasing extent. It was established that the presence of zinc ions in reactor water of atomic power plants limits the growth of the radiation dose rate from equipment; lower dose rates of gamma rays from ^{60}Co were observed. An injection of natural zinc resulted in a lower content of ^{60}Co, however, greater content of ^{65}Zn that is produced in a flux of thermal neutrons by the scheme ^{64}Zn(n, γ)→^{65}Zn. The employment of zinc depleted with respect to ^{64}Zn results in the lower cobalt contribution to the dose rate of γ-radiation (by a factor of 4–20 for various reactor types) [208].

6.3
Medicine and Biology

It has been mentioned that medicobiological investigations with isotopes and their employment in treatment of diseases are in the second place by the industrial output and demonstrate stable annual growth (by 5–10 %) in the last 30 years. A list of most of the often used stable and radioactive isotopes is given in Table 6.2.

It is worth noting that medicobiological applications make the most strict demands to the quality of isotopes. For example, for producing ^{99}Tc, which is used in more than a half of tomography isotope investigations, ^{100}Mo isotope is used with the isotopic purity of 99.99 %, which is then irradiated in cyclotron or linear accelerator. Similar demands are made on the purity of the initial products, usually tellurium isotopes in producing ^{123}I and ^{131}I, which are in the second place in the market of radioactive isotopes after ^{99}Tc.

Laser methods for obtaining highly enriched isotopes are most promising and economically profitable. One of their advantages is the flexibility of manufacturing, which is of particular importance in medicine applications.

Modern laser installations for laser isotope separation, especially by photochemical methods occupy small area, are safe in operation, and their production can be controlled by computer. In view of the already developed production of cyclotrons for obtaining radioactive isotopes for medicine they can be integrated with LIS devices and function in medical centers flexibly satisfying medical needs.

7
Appendix A: Mathematical Description of the Processes Based on Kinetic Equations

In the case of exchange reactions occurring in multi-isotopic medium, the system of kinetic equations may comprise a considerable number of equations. Hence, in an explicit form, the system of equations may become cumbersome and immense. Nevertheless, the system is substantially simplified if we employ the following procedure. Various impact processes can be presented similarly to chemical reactions:

$$A + B + C \Rightarrow D + E + F \tag{7.1}$$

We assume here that the components D, E, and F are obtained in the result of an impact between A, B, and C. If some of the components written in (7.1) are absent, then the reaction corresponds to two-particle or single-particle impacts. The components may coincide with each other. Formula (7.1) corresponds to three-particle reactions. Such reactions have not been observed in laser isotope separation yet. But in the process of induced radiation, three particles are produced, namely, two photons and atom. Three-particle impacts are theoretically possible in the presence of inert buffer gas at high pressure. Hence, for generality, the three-particle form of the reaction was accepted. Equation (7.1) can be interpreted in a more wide sense rather than a simple change in the state of particles resulting from impacts. For example, molecule deposition on a wall can be presented as $M \Rightarrow$ (with no components in the right-hand side, that is, the molecules disappear from gas medium).

We will assume that the reactions are enumerated by index m varying in the range $1 \ldots M$. The rate of mth reaction is written in the form

$$R_m = \alpha_m [A][B][C] \tag{7.2}$$

where α_m is the rate constant of the reaction; the component concentration is denoted by square brackets. If only two components are present in the left-hand side of (7.1), then we have $R_m = \alpha_m [A][B]$; if it is only one component (for example, a spontaneous decay of the level), then $R_m = \alpha_m [A]$; if there is no component, then $R_m = \alpha_m$. As an example of the latter process we may consider the passage

Laser Isotope Separation in Atomic Vapor. P. A. Bokhan, V. V. Buchanov, N. V. Fateev,
M. M. Kalugin, M. A. Kazaryan, A. M. Prokhorov, D. E. Zakrevskiĭ
Copyright © 2006 WILEY-VCH Verlag GmbH & Co. KGaA, Weinheim
ISBN: 3-527-40621-2

of atom into gas medium due to evaporation from the walls of chamber. In this case, α_m means the rate of gas intake per unit volume.

The system of equations can be expressed via the reaction parameters in a single line:

$$\frac{dn_i}{dt} = \sum_{m=1}^{M} \theta_{im} R_{im} \tag{7.3}$$

Parameter θ_{im} is defined as follows:

$$\begin{cases} 0, & \text{if } i\text{th component does not participate in the reaction} \\ -1, & \text{if } i\text{th component is present in the left-hand side of (7.1)} \\ +1, & \text{if } i\text{th component is present in the right-hand side of (7.1)} \end{cases}$$

In the framework of system (7.3), the induced radiation and the excitation of levels due to the absorption of radiation are described, however, only in the cases, where the radiation can be presented as a flux of photons considered as particles.

8
Appendix B: Operation Features of Copper-Vapor Laser Complexes

8.1
Specificity of Creating the Complexes of Copper-Vapor Lasers

Copper-vapor lasers comprised in a complex for laser isotope separation have an important function to optically pump wavelength-tunable dye lasers. For obtaining the required characteristics (the output power, spatial and temporal parameters of beam), they are usually combined into a multicascade laser system "driving generator–power amplifiers" (G+A). Certain requirements are imposed upon the construction of (G+A)-systems used as pumping sources. These are as follows.

1. Maximal possible practical efficiency at a high average radiation power with the controlled ratio of radiation power at two wavelengths of CVL: green ($\lambda = 510.6$ nm) and yellow ($\lambda = 578.8$ nm). The optimal ratio of powers in green and yellow lines is reached at a certain optimal temperature of the wall in the discharge cavity (in the range 1380–1500 °C) that provides the required density of copper vapor. Hence, the temperature of internal walls of laser tubes should be continuously controlled at the sufficiently high accuracy of $\sim 1\%$.

2. The optimal operation of (G+A)-system can only be provided at the specific mutual temporal correspondence between the instants of the radiation pulses from the driving generator and the pumping current pulses of each of the laser amplifiers. The synchronization accuracy between these signals is 1 ns. So a high accuracy is determined by the fact that the front of the current pulse, the duration of radiation pulse, the time of existence of the population inversion in the active medium of CVL, and the transit times of pulses in the active medium and between the cascades of CVL complex are close values (from several to dozens of nanoseconds). The synchronization error of a few nanoseconds results in a considerable fall in the output radiation power of the whole system. Also, the system of dye lasers is built by the same scheme "driving generator–power

amplifiers"; hence, each pulse of the pumping radiation should arrive at the corresponding dye laser amplifier synchronously with the input pulse of the proper driving generator of the system of dye lasers. At so high synchronization requirements, long operation of the whole system in the optimal regime can only be provided by the continuous control of time synchronization and shape of each radiation pulse from the driving generator and laser amplifiers with automatic adjustment of the system parameters that determine the time delays of the radiation and current pulses.

3. An important requirement to the radiation formed by the CVL complex is the high stability of the direction of the pumping beam. This is explained by the specific construction of dye lasers, namely, small dimensions of the active zone (~ 1 mm), the necessity of uniform irradiation, and so forth. In a multicascade complex, the total length of the radiation propagation tract is large (dozens of meters); hence, both slow deviations and fast fluctuations of the angular direction of beam axis may result in undesirable shifts of the pumping zone in dye chamber or in total deviation from the amplifier axis. This may cause a reduction in the output radiation power of the dye laser system or break the process of laser isotope separation. At rough adjusting errors, the dye cells may break down, because the radiation of pumping power is hundreds watt. It is obvious that the efficiency, consistency and high reliability can only be guaranteed under a continuous control and automatic adjustment of the angular direction of radiation with the accuracy of $2''$ ($\sim 10^{-5}$ rad).

The stability problem for the radiation parameters of high-power CVL systems is not sufficiently studied yet. Hence, in developing the complexes, it is necessary to investigate and make allowance for many factors influencing the stability of radiation parameters. These factors are different variants of optical schemes and the accuracy of their adjustment; engineering solutions, technologies for producing and matching separate units; combining the units into the whole system; optical and construction materials employed; an influence of mechanical vibrations and temperature regimes; pulse-periodic processes that occur in discharge plasma and the active volume of CVL, and so forth.

Let us consider, for example, how the radiation parameters of the laser systems used for LIS depend on spatial beam filtration and on the choice of the optical schemes for extracting and distributing CVL radiation between the cascades of a dye laser system (DL).

For obtaining stable parameters of the output radiation of the driving laser and amplifiers of the DL system, the pulses of the optical pumping produced by the CVL complex should be smooth and close to Gaussian (bell-shaped) in shape.

The initial optimization of the pumping pulse shape is performed at the output of the driving CVL generator by a spatial filter. By performing the angular selection, that is, by selecting from the total generator beam the radiation with the angular divergence bounded above by the size of a hole in the spatial filter, it is possible to considerably smooth the pulse shape that is usually multipeak in character, or in the limiting case, to select a single smooth pulse of a relatively short duration.

At greater number of the amplifying cascades in the CVL complex (3–4 power amplifiers and more), the smooth pulses obtained may deform again during the power amplification process. Hence, it is reasonable to continue the smoothing procedure in special optical delay lines (OD) by appropriately choosing the reflection coefficients of semitransparent mirrors of OD, splitting the initial beam of CVL and introducing the path-length differences into the optical schemes of the separated beams. In their following collimation into the optical fibers for transporting the pumping radiation to DL cells it is possible, first, to vary the duration of the initial pumping pulses and, second, to optimize their profile making it smooth bell-shaped. The delay lines in a multicascade CVL system have an important function of synchronizing the instants of the pumping pulses passing to the cells of the DL system and are placed just in front of the devices used to input radiation into optical fibers.

There are several variants of optical schemes for extracting the pumping radiation from CVL and distributing it between the cascades of a DL system. The radiation beams with the average power required for pumping the corresponding cells of a DL system are extracted from an intermediate transit of a multicascade CVL system. In another variant of the optical scheme, the beam of the driving generator is completely amplified in the CVL cascades (without intermediate extraction of radiation) and then the total output beam is split by beam-splitting mirrors into the beams of required power, which pass to the input devices for pumping the cells of DL.

There are other variants of optical schemes for the pumping systems. In particular, it seems promising to employ the injection of the generator radiation into a controlled cavity of a power preamplifier. In this case, the shape of the output radiation pulses from CVL cascades can be optimized by electronically matching the mutual time delays of electric pulses used for exciting their active media. The radiation extraction and distribution between the cells of the DL system can be made by sequential extractions from beam or by splitting the total beam at the output of the CVL system.

Thorough beam spatial interstage filtration is necessary for enhancing the contrast (reducing an influence of the amplified spontaneous luminescence) and providing the required angular divergence of the output radiation of generators and amplifiers of the whole CVL system. Let us consider a particular optical device, namely, spatial filter (SF), which, in view of the specificity of CVL (high small-signal gain and a multipeak structure of the input pulses, formed in an unstable confocal cavity of DG), has two important functions. First, SF performs the angular selection mentioned above, because it purges the radiation CVL pulses caused by the amplified spontaneous luminescence, which in CVL always precedes the main stage of radiation pulse. The amplified luminescence results in the pronounced deterioration of energy and spatial-temporal characteristics of the DL system. Second, SF is used for optimizing the shape of radiation pulses, that is, for smoothing the specific peak structure.

It is worth noting that in the cascade CVL amplification, in addition to the deformation of initially smooth pulses ("purified" by SF at the output of the driving generator), the amplified spontaneous luminescence also increases. Hence, in multicascade CVL, it is reasonable to employ SF not only at the output of the driving generator, but also between the amplifying cascades. In particular, SF should always be used at the output of CVL, directly before the system for splitting the total beam into the beams pumping the cells of the DL system.

At the typical pulse energies and average power of the output radiation of the generator its spatial filtration is not a technical problem. Nevertheless, lens air SF (two confocal lenses with a diaphragm in the common focus) conventionally used for this purpose have some drawbacks, among which two are substantial. First, a simple lens possesses chromatic aberration (the focal distances for the green and yellow pumping lines differ); combined lens elements corrected for chromatic aberration cannot be employed because the characteristics of optical glue vary under the action of the high-power radiation from the CVL generator. Second, inevitable backward Fresnel reflection from the lens surfaces results in the origin of parasitic cavities and distorts the spatial-temporal structure of the output beam. This drawback of lens SF also negatively influences the operation of the amplifier if the latter is enclosed into the cavity controlled by pulses of the driving generator according to the scheme of the input signal injection. Hence, it is desirable that the generator unit would include SF that simultaneously meets certain requirements, some of them being contradictory. The main requirements are as follows.

1. The spatial filter should be mirror-type providing "traveling wave" of radiation (that is, laser beam transportation due to successive forward-only reflections from the front surfaces of mirrors). In view of the off-axis ("breaking" the optical axis) orientation of the mirrors, SF should be corrected for spherical aberrations.

2. The spatial filter should be insensitive to chromatic aberrations.

3. The construction of SF should provide the possibility of automated, exact, two-coordinate independent adjustment of the diaphragm in the plane normal to the optical axis of the beam, and positioning the diaphragm at the desired location. Remote control of the diaphragm should be provided. The spatial filter should be evacuated to escape heating and gas convection in the focal waist region; thus, the light beam would be insensitive to the variations of the refraction coefficient caused by gas density fluctuations in the zone of convection. The SF

construction should be demountable (suitable for cleaning and replacement of elements and units) and mechanically rigid and stable. The mirror assembly should be easy to adjust. Deforming loads to SF mirrors caused by the pressure difference (atmosphere–vacuum) should be avoided. The spatial filter should be incorporated into the driving generator unit in such a way that the optical scheme, in addition to filtration and enhancing the contrast, would also convert the transversal beam dimension and match it with the input optics of the controlled cavity of the amplifier. The construction should also anticipate the possibility of compensating gradual (while passing through the optical scheme) increase of the transversal beam dimension due to the diffraction on a limited aperture. For this purpose, the scaling of beam diameter should be anticipated with the magnification $M < 1$ (for example, 0.9–0.95) by employing the corresponding mirrors (with different focal distances) and placing the diaphragm in their common focus.

The optical length (the distance between the mirrors) of SF operating at the output of the laser system based on CVL is determined by the radiation strength of the diaphragm material, corrected for minimal aberrations related to off-axis positions of mirrors, and finally amounts to 2.4–2.5 m.

As one more example, let us consider the influence of such factor affecting CVL operation as instability of the directional diagram. From experiments and the experience of working with (G+A)-systems on the basis of CVL it is known that large angular deviations of beams from the optical axis are prohibited. Such deviations lead to inevitable degradation of the most important output characteristics, namely, the average power and the angular divergence of the output radiation of a laser system. Trivial reasons for the angular deviations are mirror detuning in the cavity of a driving generator, which often results from the errors made at the construction stage (a high level of mirror vibrations caused by air cooling systems, a pulse generator, and so forth), or an accidental turn or shift of an optical element (a reflecting plate, reflecting turning mirror, SF diaphragm, and so forth).

Uncontrollable angular deviations revealing in the instability of the directional diagram (DD) of laser beams (DD trembling is more pronounced being detected in the far-field zone of CVL radiation) may also occur due to more complicated physical reasons. For example, optical nonuniformities (gradients) of the refraction coefficient may arise in the active medium of CVL related to gas-dynamic movements of the components of working mixture (fluxes of copper and neon atoms) from hot to cold parts of the active medium, thus forming domains with higher or lower density.

Also, in pulse-periodic modes of exciting the working medium of CVL by short-duration ($\sim 10^{-8}$–10^{-7} s) high-power pumping pulses, shock waves are efficiently generated due to fast energy release. These waves may cause the origin of extended spatial-periodic layers with alternate concentration shocks (compressions and depressions) of the working mixture. Under certain conditions, the result of laser radiation diffraction on such structures is the noticeable angular deflection of the beam relative to the optical axis. If the structure of such "diffraction" spatial grating is not exactly reproduced from pulse to pulse, then DD of the beam would chaotically jump within a certain range of angles in the plane normal to the optical axis. If the structure described is regularly reproduced after a certain number of pumping pulses, then DD jitter occurs according to a certain law, which may be studied experimentally. In the latter case, the control law can be determined and used for driving the actuators of the automatic stabilization system of DD in the course of operation of the CVL complex.

It is worth noting that in certain cases, for example, in using the radiation of CVL to optically pump a dye laser system, at some amplitude of the chaotic DD jumps the whole complex CVL–DL may fail to operate. Indeed, the characteristic transversal dimensions of the pumping zones in DL cascades are small and in the order of magnitude are not greater than several hundred of microns. It is obvious that chaotic mutual shifts of the pumping zones in the cascades of DL due to DD instability CVL would inevitably reduce the efficiency and, in the limiting cases, break off successive amplification of radiation power.

In addition to the above-described reason of DD jitter, one more reason is the radial thermal nonuniformity of the density of an active medium in CVL amplifiers that arises under overheating of the latter at high, exceeding the optimal value specific energy depositions into the pumping pulse. In this case, the working gas medium may be displaced from the axis of the discharge tube to its walls due to a large difference in the gas temperature at the axis and near walls, which may amount to 1500–2000 °C. At sufficiently large values of the initial gas density and the length of the working volume (1 m and longer), the active medium exhibits properties of a thermal lens with the focal length from several dozens to a few meters introducing noticeable distortions to the angular intensity distribution (that is, to the angular divergence) of the output radiation. It is reasonable to assume that such a lens would not be spherically symmetric and its curvature (that is, the density gradient of the working medium of the laser) would vary in time by certain law. The optical nonuniformity of this kind can also cause DD jitter of the CVL beam. Finally, under certain particular conditions of excitation and parameters of the working medium of the laser, optical nonuniformities arise due to the combined action of the above-mentioned and other mechanisms (for example, breaks of discharge stability).

Hence, in the general case, the instability of DD is of nontrivial nature; the features of "jitter" may be complicated and may fail to be reduced to simple laws expressed in an analytical form. This makes the choice and formulation of the control law for actuators of an automated stabilization system for DL a difficult problem for the amplifying cascades and the whole laser system. Consequently, high efficiency and operation reliability of the laser system can only be provided by continuously controlling the angular direction of laser beams.

The examples given above show that in developing a LIS complex, the complexity of systems comprised in it (in particular, CVL) should be taken into account. Also, in functioning of the LIS system it is necessary to simultaneously control (measure and regulate) a great number of parameters (laser radiation parameters and operation parameters) that influence the final efficiency of the LIS process. It is obvious that in developing such systems it is desirable to widely use mathematical simulation of separate units and the complex of LIS as a whole.

8.1.1
Specificity of Measuring Laser Radiation Parameters in CVL Complexes

Laser radiation possesses specific properties (this is also true for the radiation of pulse-periodic CVL) noticeably differing from those of ordinary optical radiation. Understanding this specificity, the correct choice of tools and measuring methods is of determining importance for the reliability and accuracy of measuring the characteristics of laser radiation. The main specific features of laser radiation are as follows.

1. High monochromaticity and coherence of laser radiation may introduce noticeable nonuniformity in the intensity distribution due to interference effects in beams deflected, for example, by plane-parallel plates. This effect can be avoided by reflecting a part of laser radiation by the plate with nonparallel faces (optical edge) so that the reflected beams propagating in different directions would not interfere. A CVL generates at two different wavelengths; hence, the main beam passing through the optical edge in addition to changing the direction (which may be inconvenient) also splits into two beams propagating in different directions due to the dispersion of the plate material. It is obvious that in the case of laser radiation, simple optical edge cannot be employed. If plane-parallel plates are used, then it is necessary to take into account possible influence of interference on the measurement results.

2. In order to reduce possible errors caused by the fluctuations of the laser beam direction at small dimensions of the photosensitive detector area (for example, in measuring the pulse duration), optical diffusers are often used (reflecting and transmitting). But in this case, speckle inhomogeneity may arise, which is the interference on a chaotic structure of the scattering material. The allowance made for the speckle structure makes it possible to correctly choose the methods and the scheme of measuring, thus providing minimal data errors.

3. Laser radiation is usually strongly polarized; in some cases, the degree of polarization is close to 100 %. In transporting polarized laser radiation through the elements of an optical system, one should remember that the Fresnel reflection coefficient considerably depends on the mutual orientation of the incident radiation and the surface of optical element if the radiation is reflected from the

boundary of two media possessing different refraction coefficients. This effect may cause serious errors in measuring absolute or relative energy parameters of laser radiation. Hence, in developing an optical system, the polarization of laser radiation should be accurately taken into account for each measuring channel. The methods calibrating a photodetector and processing the data obtained should be thoroughly elaborated.

4. A large aperture of CVL beams (the beam diameter at the output of the laser amplifier may be 60–80 mm and more) and small divergence of laser radiation, which is usually close to the diffraction limit, cause considerable difficulties in measuring far-field intensity distributions. Such measurements are conventionally made in the focal plane of the objective that focuses the beam under study. The focal length of such objective determines the dimensions of the radiation spot in the focal plane, where the analyzing multielement photodetector is placed. Usually this focal length is very long. If, for example, at the aperture of the laser beam, 60 mm, we choose the diameter of the first diffraction minimum, 1 mm, then the required focal length is about 50 m. At the focal length reduced to an acceptable level, the dimension of the diffraction image to be analyzed becomes too small. On the one hand, this results in the considerably reduced spatial resolution of the system (the typical dimension of the sensitive element in TV matrixes is about 10 μm). On the other hand, the power density at the center of the diffraction maximum becomes inadmissible. Principally, the problem can be solved by using a telephoto lens, which has small dimensions at a sufficiently short focal length. However, it is known that the smaller the telephoto lens, the more strict requirements to its quality are (the optical quality should be close to the diffraction limit). This inevitably (and considerably) increases the complexity and difficulty of its manufacturing and the cost. It is worth noting that the objective should be corrected for chromatic aberrations at both the operating wavelengths of laser radiation.

5. High power and power density of CVL, especially the peak values (in CVL, the ratio of the peak to average power may reach 2×10^4), result in the paradoxical, at first glance, requirement: the intensity of laser radiation should be substantially reduced in order to reduce the measurement errors. Indeed, the peak power of laser radiation on the sensitive area of the photodetector should not exceed the upper level of its dynamic range. Hence, the average permissible power should be lower by a factor of 2×10^4. If this is not the case, a systematic measurement error arises related to a nonlinear character of the photodetector sensitivity and to fast uncontrolled degradation of its parameters under the action of radiation with a high power density. The influence of this error can be reduced by appropriately processing the data obtained from photodetectors with software methods. Unfortunately, fast degradation of photodetector parameters requires systematic calibration and examination of the latter, which

noticeably complicates the whole construction and its operation. Furthermore, it is difficult to measure strong optical attenuation; the error of measuring the attenuation is too large, which finally results in large errors of measuring the laser radiation parameters.

The allowance made for the specific features of measuring the parameters of laser radiation considered above determines stronger requirements to the quality of the measuring system from the viewpoint of the methods, software, and hardware. Stronger requirements to measurement accuracy make it necessary to develop special methods of metrological certification, verification, calibration of sensors and the measurement system as a whole.

This work was performed in collaboration with V. V. Il'in, A. M. Kalugin, and A. B. Ugryumov. Also we are grateful to M. A. Isakov and D. N. Sitnik for technical and organizational cooperation.

9
Appendix C: Physical and Technical Problems of Increasing the Power of Copper-Vapor Lasers

Let us consider the possibilities to enhance the main parameters of CVL such as the average output power P_{las}, the repetition frequency F of pulses, and the efficiency η of converting electrical energy to radiation energy keeping the geometrical parameters of the sealed-off laser tubes unchanged [36, 48]. It is known that a copper-vapor laser with tubes 2 cm in diameter provides the specific energy of generation $\sim$20–30 µJ cm^{-3} [209] and the repetition frequency of 15–20 kHz [209]. The efficiency relative to the energy deposition is $\sim$8 % [210] in small-diameter tubes. Hence, principally it is possible to reach the average linear power exceeding 100 W m^{-1} at the efficiency of 3 % and greater obtained at low average levels of generation [210, 211].

Generation in vapor–gas mixture Cu–Ne occurs in the transitions $^2P^0_{3/2}-{}^2D_{5/2}$ at $\lambda = 510.6$ nm and $^2P^0_{1/2}-{}^2D_{5/2}$ at $\lambda = 578.2$ nm (see Fig. 9.1).

The levels $^2P^0_{3/2}$ and $^2P^0_{1/2}$ are efficiently excited from the ground state $^2S_{1/2}$ by electron impact because these resonance levels are strongly bound to the fundamental allowed electric dipole transition. The transitions $^2D_{3/2;5/2}-{}^2S_{1/2}$ are forbidden. Copper atoms may rapidly relax from the states $^2P^0_{3/2}$ and $^2P^0_{1/2}$ to the ground state $^2S_{1/2}$ via spontaneous emission; although it does not occur due to reabsorption of radiation under laser operating conditions. The state ^2D relaxes due to deactivation on walls and electron de-excitation. Hence, the population inversion is created on the transitions $^2P^0_{3/2} \rightarrow {}^2D_{5/2}$ and $^2P^0_{1/2} \rightarrow {}^2D_{3/2}$ and the laser generates in a self-terminating mode due to faster pumping of the upper level as compared to the lower level (until the moment, when $N_r \sim (g_r/g_{ms})N_{ms}$, where N_r is the population of resonance state (RS); N_{ms} is the population of metastable state (MS); g_r and g_{ms} are the statistical weights of RS and MS, respectively).

The problem of increasing P_{las} at the sacrifice of frequency F at a high generation efficiency η leads us to discuss the problem of limiting the frequency–energy characteristics (FEC). As early as in the first investigations of CVL and of other lasers on self-terminating transitions, the assumption was postulated that F is determined by the departure of atoms from the lower MS states to walls. Nevertheless, direct experiments with CVL operating at $F > 10$ kHz show that there

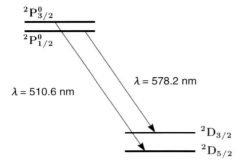

Fig. 9.1 Low energy levels of copper atom.

exist other faster de-excitation processes that provide high repetition frequency of generation pulses.

Presently, FEC limitation of lasers on self-terminating transitions, in particular CVL, is usually related to insufficient relaxation of population of MS states [212] and to slow recombination in plasma afterglow [213]. In the former case, the pre-pulse MS population (N_{ms0}) may be so high that it affects the energy of generation pulse at higher F and then reduces P_{las}. At a certain threshold value (N_{ms}^{thr}) it becomes so high that the generation fails to start during the pumping pulse because we have $N_r - g_r/g_{ms} N_{ms} < 0$.

In the second case, at closer pulses and higher pre-pulse concentration of electrons (N_{e0}), various processes occur, which hinder in obtaining generation. The influence of N_{e0} is implicitly or explicitly taken into account in the kinetics of pulsed-periodic lasers via the fall in the rate of electron gas heating [214–217], skin-effect [218], step depopulation processes [219], the increase in the degree of vapor ionization, and so forth.

It is obvious that at various operation conditions of gas-discharge CVL, each of the mechanisms may be revealed and the uncertainty arise only in specifying the ranges of particular medium parameters and pumping, in which one or another mechanism of FEC limiting prevails. Let us determine the physical–technical operation conditions and the requirements for laser tubes and excitation systems in order to obtain high P_{las} and η parameters for gas-discharge tubes [36, 48].

Uncertainty in understanding the role of N_{ms0}, N_{e0} and their influence on the energy characteristics of CVL leads to contradictory results in works of different authors. According to [220], for example, limitations related to N_{ms0} in a tube with the diameter $d = 2$ cm are observed at $P_{las} \sim 15$ W m^{-1}. Nevertheless, in commercial CVL with the tube of the same diameter we have $P_{las} > 40$ W m^{-1}

[36,48]. This contradiction is the demonstration of the fact that the channels of MS relaxation in [48] and [220] are different and depend on the excitation conditions.

It can be shown that in tubes with $d \sim 2$ cm and with the linear generation power ~ 50 W m^{-1}, the pre-pulse concentration of MS is negligible and has no effect on the output power; the latter fact makes it possible to obtain the generation efficiency greater than 1%. At lower efficiency caused by a mismatch between the pumping generator and a laser tube, the afterglow population of MS noticeably increases and N_{ms0} may become the main factor reducing the generation power in a pulse-periodic regime.

The quality of matching, for relative measure of which we will take the efficiency of generation really achieved in the optimal conditions, strongly influences the physical processes that occur during the pumping pulse and in afterglow. Without taking it into account, it is impossible to uniquely interpret the mechanism limiting P_{las} in high-power CVL and to determine the ways of increasing it. The quality of matching strongly influences the behavior of N_{ms} in afterglow of Cu–Ne mixture used in sealed-off lasers.

Let us consider the influence of the pre-pulse concentration of metastable states on the generation properties of optimized sealed off-CVL.

For the first time, thesis on the negligible influence of N_{ms0} on the limitation of P_{las} in CVL was formulated in [213]. The results [213] were obtained for optimized lasers [211] at moderate pumping levels ($P_{pump} \sim 450$ W m^{-1}) and output power ($P_{las} \sim 7$ W m^{-1}). Modern lasers, including sealed off lasers operate at much higher powers $P_{pump} \sim 4$ kW m^{-1} and $P_{las} \sim 40$ W m^{-1} [48]. Data from [220] show that the main channel of limiting P_{las} in such conditions is high N_{ms0}. These conclusions contradict to the results presented in [221] for CVL and in [222] for laser on copper bromide, where the main limiting channel is a high value of N_{e0}. In order to clarify the reasons of the discrepancy, the influence of N_{ms0} on generation characteristics of high-power CVL was studied [223] with the help of the laser complex described in [50] and in Chapter 2.4. The radiation of the driving CVL generator of the type GL-201 with an unstable cavity was directed to two tubes of the type "Crystal LT-40Cu" placed in series [48]. The start of the pumping generator of the second tube could be shifted in time relative to the synchronously operating pumping generator of the first tube and the driving generator in the limits from 0 to ±90 µs (the interval between pulses).

The typical output power of a "Crystal LT-40Cu" single tube in the mode of generator–amplifier exceeded 50 W m^{-1} at the total power of up to 70 W, and the generation efficiency (relative to the power consumed from rectifier) was 1.3–1.8%, which is close to the efficiency at small pumping power [211,213]. The typical pumping parameters, at which the maximal generation power is observed are: the repetition frequency $F = 11$ kHz; the voltage amplitude $U_a \approx 25$ kV; the rise front duration ~ 25 ns; the half-height current duration ~ 45 ns; and the average consumed power is up to 4.5 kW. The pumping generator was made by the scheme of partial discharge through a GMI-29A tube. The excitation pulse from the generator to the tube was transmitted through a cable with the impedance 75 Ω 10 m in length. Such a driver could not provide the quality of matching between the driving generator and a load close to that in [211], hence, at the maximal generation power the amplitude of the first positive current spike, arising in twice the time of pulse passing through the cable was ∼50 % (as compared to 5 % in [211]).

The population of MS and its influence on the generation characteristics were determined by using the method of separating the domains of absorption and generation developed in [222]. The advantage of this method is the undistorted revealing of N_{ms0} influence on the generation characteristics of lasers on self-terminating transitions separately from the influence of N_{e0}.

The method of the experiment is as follows. The pumping level is adjusted so that the generation power at the output of the first tube is ∼1/2 of that of the second tube under an ideal synchronization of exciting pulses. In other words, the generation characteristics of laser tubes were made identical. Then, the pumping of the first tube was retarded and the power of its radiation, passed through the second tube, was measured. The influence of superluminescence from the second tube was avoided by carrying out the power measurements in far-field zone.

In Fig. 9.2, curves 1–3 show the time dependence of power attenuation for the first tube at the output powers of 55 W, 35 W, and 27 W, respectively. Curves 1′–3′ show, correspondingly, the decay of the total population of Cu(^{2}D) in afterglow. One can see that at a given repetition frequency and the generation conditions

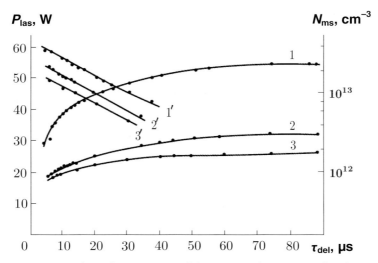

Fig. 9.2 Time dependence (1, 2, 3) of the power radiation passed and (1′, 2′, 3′) the population of Cu(^{2}D) in afterglow at the output power of 55 W, 35 W, and 27 W, respectively.

that are close to optimal ($P_{las} = 55$ W, $\eta = 1.31\%$), the pre-pulse population of MS has no pronounced influence on the generation parameters of CVL.

The result obtained does not contradict to the data from [224, 225], according to which in typical excitation conditions the time of MS depopulation for Cu atom is $\tau = 1/k_e N_e < 0.5$ µs (where k_e is the rate of de-excitation by electrons and N_e is electron concentration). The experimentally found relaxation times are considerably longer than 0.5 µs. Hence, the time of relaxation of MS is determined, similar to [225], by the rate of electron cooling in the considered time interval.

It is obvious that the shape of the pumping pulse and the accuracy of matching between the generator and the laser tube affect the generation efficiency and the relaxation of MS population.

It may be shown that the matching between the pumping generator and the laser tube substantially affect the mechanism of relaxation in short-range and long-range afterglow. At unsatisfactory matching, which is typical for discharge-heating laser tubes, the current fluctuations may last for 100 ns [226, 227] and even a few microsecond [228]. Electron temperature T_e sustained at high level results in a considerably longer relaxation time of MS in afterglow, which rises N_{ms} and N_e because of the useless energy contribution to plasma. This makes the decay of MS population longer, the influence of MS on the generation power becomes noticeable, and specific features are revealed in the limiting mechanism of each particular case.

One forced reason that leads to strong mismatch is a nonoptimal circuit of laser excitation. In [211], in an example of CVL it was shown that the optimal value of the storage capacitor C in the conditions of maximal efficiency is

$$C_{opt} \sim 15 d^2 / l \tag{9.1}$$

where d and l are the diameter and length of laser tube, respectively, (in cm) and C is measured in nF. The fulfillment of this condition made it possible to obtain 3% efficiency of CVL in the single-pulse mode and 2.6% in the pulse-periodic operation. Excursion from the optimal capacitance and from the energy related with it influences the laser efficiency and, most noticeably, the behavior of MS population in afterglow.

Let us consider six cases covering a wide range of various pumping conditions by the geometrical dimensions of laser tubes (see Table 9.1). In five of these cases the specific radiation energy is almost the same. In Fig. 9.3, the corresponding MS decay curves are shown for Cu($^2D_{5/2}$) afterglow. The data on MS decay under the conditions given in [213] are presented in [229].

As it follows from Table 9.1 and Fig. 9.1, the better laser is matched by the parameter C with C_{opt}, the higher its efficiency and lower afterglow concentration of MS are, independently of the diameter of tubes. A comparison of data from [220, 231] with those given in [45, 213, 229] shows that at practically the same energy extraction the differences in MS populations are of two orders of magnitude. They are

Table 9.1 Laser characteristics at optimal operating temperatures

ω_{pump} (mJ cm^{-3})	1.6	1.0	0.35	0.23	0.2	1
ω_{las} (µJ cm^{-3})	4.0	3.34	2.7	2.5	3.2	13.1
η (%)	0.25	0.33	0.76	1.07	1.6	1.31
C_{opt} (nF)	—	0.54	6.4	4.6	4.4	—
C (nF)	*	6.0	12	10	4	*
C/C_{opt}	—	11.1	1.9	2.2	0.9	—
d (cm)	2.0	1.8	8.0	8.0	2.7	2
l (cm)	40	90	150	210	25	120
P_{pump} (kW)	2.0	1.8	13.2	12.1	0.11	4.2
P_{las} (W)	5	6	100	130	1.8	55
F (kHz)	10	7.85	5.0	5.0	4.0	11.1
δ	~1	0.72	< 0.12	< 0.09	0.02	< 0.02
Reference	[220]	[231]	[230]	[45]	[213]	[223]

ω_{pump} and ω_{las} are the specific energies of pumping and generation, respectively.
η is the laser efficiency.
C is the working capacitance.
P_{pump} and P_{las} are the average powers of pumping and generation.
F is the pulse repetition frequency.
* The capacity is partially discharged through a switch.

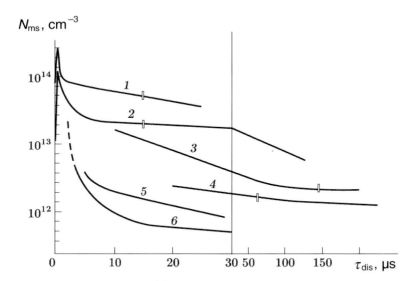

Fig. 9.3 Population of Cu(4s^2 ^{2}D) level versus afterglow time.
(1) results from [220]; (2) results from [231]; (3) results from [230];
(4) results from [45]; (5) results from [223]; (6) results from [213].
The delay of the generation is marked with | . Vertical line separates different time scales. Dashed part in curve 6 refers to extrapolated results.

caused by small variations in the specific pumping energy (see [213, 220]; Table 9.1 and Fig. 9.3), which, however, result in a pronounced difference in the value of N_{ms} that arises in 10–15 μs after the pumping pulse stops. For better understanding the mechanism of retarding the decay of MS in nonoptimized lasers, let us consider the balance of electron energy during the action of pumping pulse and in afterglow.

Resonance states are intensively populated during the active phase of the pumping pulse, which is conditionally limited by the instant at which the generation terminates. During this period, step processes are efficient [219, 232], which lead to considerable population of the states higher than the resonance states [233, 234] and to the ionization of copper atoms. Metastable states have too little time to be populated to a noticeable degree. According to the theoretical [215, 235] and experimental [228, 236] data, the generation phase terminates at $N_e \sim (2\text{–}5) \times 10^{13}$ cm^{-3} at the moment $\tau_f \sim 100$ ns after the start of current pulse. Hence, the maximal density of atoms passing to MS through the direct excitation channel (Cu(4s ^{2}S) + e → Cu(4s^2 ^{2}D) + e $-\Delta E$) is

$$N_{ms}^f(^2\text{D}) = \int_0^{\tau_f} \langle \sigma v \rangle N_a N_e dt \leq 10^{13} \text{ cm}^{-3} \tag{9.2}$$

where $\langle \sigma v \rangle$ is the rate constant of MS excitation taken from [235, 237] and N_a is the concentration of copper atoms. Approximately the same number of atoms pass to metastable states through the generation channel at the energy extraction $\omega_{las} \sim 4$ μJ cm^{-3} (see Table 9.1), which is confirmed by calculation [237]. It is obvious that at the moment of generation termination we have $N_r \sim (g_r/g_{ms}) N_{ms}$.

If electrons are not heated by following current waves at the end of this phase (as it is in the case of the optimized lasers matched with the pumping generator), then T_e rapidly falls due to the step processes of the type Cu(4p ^{2}P) + e → Cu(ns ^{2}S) + e $- \Delta E$. According to the data from [235, 238], the rate of the step processes $k_e^\times$ at $T_e > 1$ eV exceeds 10^{-7} cm^{3}s^{-1} at the total population of resonance and other highly excited states $\sum_k N_k \sim 10^{14}$ cm^{-3}. This results in the characteristic time of electron cooling $\sim 10^{-7}$ s. Simultaneously with the fall of T_e, the populations of all levels are depleted, including MS with its de-excitation rate constant [224, 225] to the values, at which the step processes become inefficient. The last of the processes that are important for electron cooling terminates the process Cu(4p ^{2}P) + e → Cu(5s ^{2}S) + e $- \Delta E$ (depopulation of MS to the nearest highly excited state possessing the dipole coupling with MS). This occurs at $T_e \sim 0.4$ eV, where the rate of step reactions becomes equal to the rate of MS de-excitation that warms up electrons. If the parameters N_r and N_{ph} (N_{ph} is the specific number of emitted photons) are comparable with N_{ms}^f (see equation 9.2), then such mechanism of electron cooling along with the de-excitation of MS to the ground state (Cu(4s^2 ^{2}D) + e → Cu(4s^2 ^{2}S) + e −ΔE) results in the fall of N_{ms} down to less than 10^{13} cm^{-3} in short-range afterglow (the phase of fast reduction of N_{ms}, where the population cannot follow the fall of T_e) with the following freezing (the phase of slow reduction of N_{ms}, where the population of MS is determined by the relaxation of T_e).

If electrons are heated by subsequent waves of current, then, according to [235] (with the operation conditions of [231]) we have $T_e > 1$ eV for as long as the afterglow duration $\tau_{dis} \sim 500$ ns and T_e remains high (greater than 0.5 eV) for several microseconds [218, 239, 240]. Up to this moment, MS are populated to $N_{ms} \sim 10^{14}$ cm^{-3} [237] and higher [220, 231], whereas all highly excited states, including MS, are de-excited to the level, at which $\sum_k N_k \ll N_{ms}$. Hence, the only remaining channels of electron cooling are ambipolar diffusion (in small-diameter tubes at low pressure of buffer gas) and elastic collisions with heavy particles [241] (de-excitation of MS in copper atoms in collisions with heavy particles and, hence, the cooling in this channel are inefficient [225]).

The time constant of electron cooling in collisions with neon under the optimal conditions (at the pressure of neon $p_{Ne} \sim (20-30)$ Torr) is 20 μs. This value is typical for the second phase of the decay of MS population in both optimized and nonoptimized lasers. Nevertheless, this phase starts at substantially different N_{ms}, which noticeably influences the relaxation of MS and the recovery of generation in the subsequent pulse (see Fig. 9.3).

The exponential character of the reduction of N_e in afterglow [242] is kept at noticeably different energy depositions and different initial values of N_e. Hence, the conditions with respect to N_e are restored in optimized and nonoptimized lasers with the approximately equal delays. Nevertheless, in optimized lasers, the fast drop of T_e terminates at $N_{ms} < 10^{13}$ cm^{-3}; as fast as in a few microseconds, medium restores the ability to generate with respect to criterion N_{ms0}, independently of the working medium of laser, the pumping method, and the method of measuring N_{ms} [223, 243, 244]. In this case, the dominating influence of N_{e0} on the recovery of generation can be easily detected. In nonoptimized lasers, the reduction of N_{ms} to the level less than 10^{13} cm^{-3} occurs much later, and it may found that N_{ms0} would not only influence the generation energy but also entirely determine it [220].

We may estimate the influence of N_{ms0} for the cases presented in Table 9.1. Up to now, no convenient methods have been elaborated for estimating its influence on the generation characteristics of self-terminated lasers. The qualitative methods developed, for example, in [245, 246] are not satisfactory, because they contradict to the quantitative calculations of the same authors. The contradictions are displayed in that limitations occur at substantially different populations of MS [20, 246]. For example, in [220], generation ceases at $N_{ms0} \sim 4 \times 10^{13}$ cm^{-3}, whereas in [246] it occurs at $N_{ms0} \approx 1.5 \cdot 10^{12}$ cm^{-3} at close energy extraction at the operation repetition frequency. The qualitative methods based on computer simulations of laser kinetics [215, 221, 235] are also vulnerable because of the inaccurate determination of the rate constants of numerous important processes. Hence, in estimating the influence of MS on the generation parameters, it suffice to use semiquantitative calculation based on the directly measured parameters, such as the specific energy extraction, the absorption coefficient prior to the pumping pulse start, and so forth.

The method is based on the model of laser generation with absorbing particles inside the cavity. Their presence leads to weaker generation. If the particles in the low state do not relax during the generation pulse (it is the case of laser on self-terminating transitions), then their presence reduces the generation energy, as compared to free generation, by the value $\Delta \omega_l = h\nu N_{abs}/2$ [247], where N_{abs} is the concentration of absorbing atoms. The geometrical positions of absorbing particles in a cavity and the character of populating the working levels during the pumping pulse are not important. Starting from this model, it is possible to write the relative attenuation of the generation of self-terminated laser caused by the pre-pulse population of MS in the form

$$\frac{\Delta W}{W} = \left(\frac{g_r}{g_r + g_{ms}}\right) \frac{N_{ms0}}{N_{ph0}} \tag{9.3}$$

where N_{ph0} is the specific number of photons emitted at $N_{ms0} = 0$. Under the pulse-periodic pumping, N_{ph0} has meaning of the specific number of photons for the operation mode, where N_{ph} is independent of F. Under the threshold conditions, $\Delta W/W = 1$, the equality

$$N_{ms}^{thr} = \left(\frac{g_r + g_{ms}}{g_r}\right) N_{ph0} \tag{9.4}$$

holds.

In experiments, the source of information for calculating N_{ms} is often the measurement of the absorption coefficient k_0^- for the working transition. It is easy to show that

$$\Delta W/W = k_0^- / k_{max}^+ \tag{9.5}$$

where k_{max}^+ is gain at the maximum inversion. For CVL, k_{max}^+ can be calculated with sufficient accuracy from the measured energy extraction ω_{las}. Since we have

$$\omega_{las} = N_{ph} h\nu = \frac{N_r^{max} - (g_r/g_{ms}) N_{ms}^{max}}{1 + g_r/g_{ms}} h\nu \tag{9.6}$$

and the absorption (or amplification) cross-section for both generation lines weakly depends on temperature in a wide range of conditions [32], we obtain

$$\Delta W/W = k_0^- / (7.6 \cdot 10^{-2} \omega_{las}) \quad \text{for green line} \tag{9.7}$$
$$\Delta W/W = k_0^- / (5.6 \cdot 10^{-2} \omega_{las}) \quad \text{for yellow line} \tag{9.8}$$

where $h\nu$ is the quantum energy and ω_{las} is the specific radiation energy measured in µJ cm^{-3}.

If the generation weakening and termination at more frequent pumping pulses follow the conditions, in which formulae (9.2) to (9.8) have been derived, then these effects are caused by the pre-pulse concentration of MS. If these conditions are not fulfilled, however, the generation weakens (that is $\Delta W/W > k_0^-/k_{max}^+$), then it is obvious that the generation weakens due to other processes not connected with N_{ms0}. In this case, the relation

$$\frac{k_0^-}{k_{max}^+}\left(\frac{\Delta W}{W}\right)_{real}^{-1} = \left(\frac{g_r}{g_r+g_{ms}}\right)\frac{N_{ms0}}{N_{ph0}}\left(\frac{\Delta W}{W}\right)_{real}^{-1} = \delta \quad (9.9)$$

determines the degree of N_{ms0} influence on the real attenuation of generation $(\Delta W/W)_{real}$.

The part β of participation of the processes not connected with N_{ms0} can be estimated by the relationship

$$\beta = 1 - \frac{k_0^-/k_{max}^+}{(\Delta W/W)_{real}} \quad (9.10)$$

In the threshold conditions, we have $\beta = 1 - k_0^-/k_{max}^+$. The parameter β obtained in this way determines the low limit of the part of the processes that reduce, in addition to N_{ms0}, the output energy. This is explained by the fact that in real lasers with self-terminating transition, the growth of the specific radiation energy is not always proportional to k_{max}^+. It may be higher, because N_r^{max} and k_{max}^+ fall due to superfluorescence, including the case of nonaxial beams [248]. In the generation regimes, the negative effect of superfluorescence is reduced or even canceled, which causes more rapid rise of W as compared to k_{max}^+. In other words, the estimate by formulae (9.5), (9.7), and (9.8) yields the upper limit of the degree of influence of the pre-pulse MS concentration on laser characteristics. Formula (9.3) is exact, but N_{ms} is not the initially measured parameter of laser. Additional a priori information is necessary for calculating N_{ms}, for example, the transition probability, the profile of the absorption line, and so forth.

In view of formulae (9.3) and (9.9), let us estimate the role of N_{ms0} in the termination of generation under the conditions given in Table 9.1. From the character of the dependence of the average generation power P_{las} or of the pulse energy on the repetition frequency F (period T) (see Fig. 9.4 and data from [218, 224, 249]) one may conclude that $N_{ph0} \sim 1.7 N_{ph}^{max}$, where N_{ph}^{max} is the number of photons emitted in the conditions of maximal P_{las}.

The relative part δ responsible for the participation of N_{ms0} in the generation weakening was calculated in the above-mentioned conditions and is presented in Table 9.1. One can see that only in [220] and [231] the threshold conditions are mainly determined by the influence of N_{ms}. Nevertheless, the authors of these works were most farther from the optimal conditions. In all the other cases from

9 Appendix C: Physical and Technical Problems of Increasing the Power of Copper-Vapor Lasers

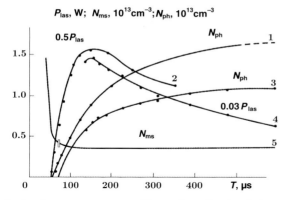

Fig. 9.4 Average power, specific number of emitted photons, and the population level Cu$(4s^2\,^2D_{5/2})$ versus the period between pulses. The delay of the generation start is marked with | sign. Curves 1 and 2 are results from [224]; curves 3, 4, 5 result from [218].

Table 9.1, similar to [218] (see Fig. 9.4), the generation weakening and termination occur well before it follows from relationships (9.5) to (9.9). In those cases, we have $\delta < 0.15$. Hence, the main reason limiting the optimal repetition frequency of pulses and the maximum generation power is, similar to [225], the growth of N_{e0} at higher F. This growth leads, in particular, to the relative increase of the rate of MS excitation at the front edge of current pulse. As an example, in Table 9.2 the experimental data from [229] are presented for the rate of exciting RS and MS at the initial stage of generation pulse at $F = 2.8$ kHz and 12.5 kHz in the conditions of [213].

Calculation of the influence of N_{e0}, based on the models from [214, 215] with the allowance made for the rates of exciting working levels [235], de-excitation rates for MS [225], and the rates of plasma recombination in the mixture Cu–Ne at increased pressure of Ne [250] yields the dependence of the relative generation energy on the time interval between pulses shown in Fig. 9.5 as curve 1. The dependence of $P_{las}(1/T)$ was calculated with the allowance made for N_{e0} only (curve 2), for N_{ms0} only (curve 3) (see curve 1 in Fig. 9.1), for the combined influence of N_{e0} and N_{ms0} (curve 4). Curve 5 corresponds to the linear growth of P_{las}

Table 9.2 Redistribution of the rates of exciting upper $(dN/dt)_r$ and lower $(dN/dt)_{ms}$ working levels at various repetition frequencies of pulses (Δt is the time, in which N_r and N_{ms} change by the value 4×10^{12} cm^{-3})

F (kHz)	$(dN/dt)\Delta t$ (10^{12} cm^{-3})	Δt_r (ns)	$(dN/dt)_r$ (10^{20} cm^{-3}s^{-1})	Δt_{ms} (ns)	$(dN/dt)_{ms}$ (10^{20} cm^{-3}s^{-1})
2.8	4.0	3.4	11.8	7.2	5.6
12.5	4.0	4.5	8.9	6.5	6.2

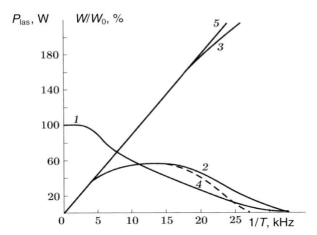

Fig. 9.5 Dependencies on the time interval between pulses.
(1) relative generation energy; (2) generation power calculated taking into account N_{e0}; (3) generation power calculated taking into account N_{ms0}; (4) generation power calculated taking into account N_{e0} and N_{ms0}; (5) linear growth of P_{las} with the repetition frequency of pulses.

with $1/T$. One can see that there is a substantial reserve for increasing the power of sealed-off CVL tubes with a diameter $d = 2$ cm by overcoming the limitations connected with N_{e0}.

Greater pressure of Ne [250], introducing hydrogen [213] and electronegative additives [251] reduce N_{e0} and increase the resistance of gas-discharge tube, which enhances its matching with the pumping generator. This facilitates the heating of electrons by the rising edge of current pulse, thus overcoming the limitations connected with N_{e0}. Also, the current oscillations following the pumping pulse are reduced, which weakens the influence of N_{ms0}.

Let us compare the results obtained in [223] on the decay of N_{ms} in high-pressure lasers (see Fig. 9.2) with the data obtained at low pressure of neon (see Table 9.1 and Fig. 9.3). One can see that at the efficiency close to the case of [213] (see Table 2.1), the specific energy deposition is much greater, which results in more than fourfold energy extraction. A comparison of the optimal operating temperatures [213] and laser tubes [48] shows that the operating concentration of copper atoms increases in approximately the same degree. The population of Cu($^2D_{5/2}$) in afterglow, normalized to energy extraction $\omega_{las} = 3.2$ μJ cm^{-3} was shown in Fig. 9.3 for the tube of type "Crystal LT-40Cu" [223]. One can see that the curve of MS population in afterglow resides under the population curve corresponding to $\eta = 1.07\%$ [45], however, higher than the curve corresponding to $\eta = 1.6\%$ [213]. This fact confirms the conclusion made above about the increase of N_{ms} in afterglow at lower generation efficiency. Nevertheless, in the case, corresponding to

Fig. 9.2 at the pulse repetition frequency used and the generation conditions close to optimal ($P_{las} = 55$ W), the pre-pulse population of MS has no influence on the parameters of CVL generation. The identical characters of the decay of N_{ms} (see curves 1′–3′ in Fig. 9.2) and identical current pulses prove that in afterglow N_{ms} is determined by the value T_e, at which it freezes. The latter, in turn, depends on the degree of matching between the pumping generator and the laser tube. It is seen in Fig. 9.2 that although the matching is not perfect (it is seen from current oscillograms), N_{ms0} does not determine the generation parameters of tubes of the type "Crystal LT-40Cu."

Hence, the influence of matching [211] between the pumping generator and the laser tube on the relaxation of MS in afterglow and the effect of the latter on the efficiency and FEC of high-power CVL are of principal importance. The value of the maximal population of MS and following relaxation in afterglow strongly depend on the degree of matching between the pumping generator and the laser tube and, correspondingly, on the laser efficiency. This leads to differentiation of the mechanisms limiting the frequency-energy characteristics of laser. At high laser efficiency, provided by good matching, the main factor limiting the growth of the output power at higher repetition frequency is high pre-pulse concentration of electrons. At unsatisfactory matching and low efficiency, in certain cases the main reason of the limitation may be high pre-pulse concentration of MS. In further removing the limiting mechanisms mentioned above, other factors of thermal nature (caused by overheating of active medium, especially at the center of the latter) may become foremost. The ways of avoiding overheating are of engineer character and are gradually verified with other laser types.

There is no doubt that by further optimizing the excitation conditions, which neutralize the influence of N_{e0}, namely, by increasing the neon pressure accompanied with the corresponding rise of the working voltage, introducing hydrogen, making the leading front of voltage and current pulses shorter, it is possible to considerably increase both the average generation power of tube of the type "Crystal LT-40Cu" and their efficiency.

Let us consider some of the conditions mentioned.

Increasing the pressure of active medium is not trivial task not only for CVL but also for all gas lasers. The problem is that at pressures exceeding certain critical value (dependent on gas type, current, and so forth), the state of positive column of self-sustained discharge becomes unstable. After some time after the discharge ignition it contracts. Positive column contracts into current pinch, in which the plasma parameters (the degree of ionization, gas and electron temperatures) considerably differ from those in conventional glow discharge. This makes it useless for obtaining generation in active medium. Not considering problems on the roles of particular contraction mechanisms, it is worth noting that the nonuniformities of a gas-discharge plasma that arise in volume and in near-electrode domains in the direction normal to that of electric field substantially hinder the stability of discharge burning. Hence, the contraction problem for high-pressure discharges, especially in large volumes, still remains actual. Nevertheless, it is proved that in large-diameter CVL, the nanosecond pulse-periodic discharge remains uniform [253]. Investigations show [254–256] that for lasers on metal vapors it is possible to suggest the following interrelated mechanisms stabilizing the discharge at considerable pressures of the vapor–gas active media: the volume character of producing electrons due to the preferable ionization of metal vapor at the limited concentration of the latter; the incomplete deionization of plasma in pulse-periodic discharges, which provides the pre-pulse electron density exceeding the level needed for initiating volume discharge.

The essence of the first mechanism is that every local increase of N_e due to the developing ionization-overheating instability occurs at the sacrifice of metal vapor ionization, since the potential of even double ionization is far lower than the ionization potential of the buffer gas. The volume character of metal distribution in a gas-discharge tube and its limited concentration determine additional uniform increase in the electron density.

The second mechanism is determined by the fact that the method of exciting vapor-gas mixture itself may be the stabilizing factor, for example, the pulse-periodic discharge. The repetition mode of operation results in that plasma has no enough time to recombine totally during the time lapse between pulses. As it was shown, the value of the pre-pulse electron concentration in CVL may be $N_{e0} \approx 10^{12}$ cm^{-3}. Hence, the pumping pulse stimulates the evolution of current in the already conducting channel due to the residual conductivity. Actually the residual electron density is just N_e related to preionization, which is much higher than the value needed for forming volume discharge in pure gases. It is worth noting that at higher pressure of the buffer gas N_e should increase, and the needed breakdown electron concentration at the fixed parameter E/N is inversely proportional to the cube gas density. Hence, it is necessary to increase the concentration of metal atoms at higher pressure of buffer gas.

At the ignition stage of discharge burning in gas–vapor mixtures, single or double ionization of metal atom by electron impact occurs followed by ionization of buffer gas atoms. During the current pulse, the electron density uniformly increases. After pulse termination in the steady-state pulse-periodic discharge, N_e falls to certain value, which, to the instant of the next pumping pulse, is higher than the level needed for initiating the volume discharge. Actually it is possible to choose the range of contents of the components (metal vapor–buffer gas) for the discharge in high-pressure gas-vapor mixture, which would provide the self-stabilization of the gas discharge and obtaining volume pulse-periodic discharge at pressures considerably higher than atmospheric.

Gas-discharge plasma in lasers on metal vapors operating at high pressure is specific in that the initial conditions should provide the burning of uniform discharge. On this background, the conditions needed for generation itself should remain or arise. It is obvious that, generally speaking, it may not be the case. High pressure deforms the distribution functions of electrons over energy, thus influencing all the processes, in which electrons are involved (excitation, de-excitation, ionization, recombination, and so forth). The pressure of buffer gas affects such processes as diffusion, recombination, and the elementary reactions involving atoms of inert gas. It also influences the ratio of rate constants for elastic and diffusion cooling of electron gas in the period between pumping pulses and, hence, the relaxation rate of plasma parameters. The resistance of the discharge gap also varies as well as the conditions of energy deposition into the active medium (the energy part to gas heating increases, the ratio between active and inductive impedances of load varies, and so forth).

We may conclude that from the viewpoint of physics of gas discharge, the increase of pressure in active medium is not an insurmountable obstacle limiting the development capabilities of CVL. Moreover, in other laser types, for example, in lasers on vapors of ionized europium (He–EuII) under the condition of remaining the ratio E/P constant, many times the optimal pressure exceeds atmospheric pressure (by the factor of five) at high uniformity of active medium [254].

From viewpoint of keeping the generation conditions in Cu–Ne active medium, it is worth noting works [48, 256–258], in which gas-discharge CVL with the longitudinal pumping demonstrated the capability of generating at pressures close to atmospheric and higher.

The introduction of hydrogen into active medium of laser on vapors of copper and its salts [85, 86, 213, 222, 259, 260] results in greater P_{las} due to faster recombination of plasma in gas-discharge afterglow. Presently, the addition of hydrogen into commercial lasers is a customary practice. In sealed-off lasers of series "Crystal" with the hydrogen addition of 1–14 Torr at the neon operating pressure of 150–250 Torr, P_{las} and η increased by a factor of 1.5 [260]. Theoretical models for lasers with addition of hydrogen are also known (see, for example [261]).

The requirements for the excitation conditions of CVL active medium can be formulated as follows.

For efficient pumping of active medium, the exciting voltage pulse should have a sharp edge on the active component of the impedance of the gas-discharge tube. It is necessary because the electron temperature T_e that determines the rates of populating laser levels follows the variations of the electric field intensity across plasma [32, 233]. The slow rise of electron temperature initially determines the population of low energy levels until $T_e < 1.7$ eV, which is a substantial factor limiting FEC [209, 211].

The duration of excitation pulse should be comparable with the time of the population inversion existence, because it determines the energy deposition to active medium per single pulse and the population of MS in afterglow.

Based on the above consideration, we may formulate the main requirements for the development of laser power supply systems and for the choice of switches forming the excitation pulses. Usually, a storage capacitor is discharged through a thyratron for pumping the active medium of CVL. The fast leading edge of the voltage pulse on the active component of the tube impedance is determined by an aperiodicity condition for the discharge circuit in the exciting schemes with the partial discharge of storage capacitor and by the frequency of free circuit oscillations in the schemes with the complete discharge [262, 263]. The leading edge is noticeably influenced by switches [263] and by the processes that occur on the electrodes of gas-discharge tube and in near-electrode cold zones [264]. The condition of an efficient pumping is also the formation of the pumping pulse duration comparable with the time of the population inversion existence. This requirement determines the necessity of perfect matching between the source and the load [223], which may be difficult to fulfill, because the Q-factor of the discharge circuit is high at the end of the exciting pulse and the energy dissipation stored in the reactive component is of oscillating character.

Let us formulate the main criteria and conditions for obtaining high generation power at a substantial efficiency based on numerous works devoted to study of CVL.

1. The duration of the pumping pulse should be comparable with the time of the population inversion existence.
2. The rise time of the voltage pulse should be short for less population of the lower levels at the initial stage of the discharge.
3. It is desirable to provide aperiodic character of gas discharge.
4. The laser tube should be perfectly matched with the power supply.
5. The pressure of active medium should be made higher with correspondingly increasing the voltage and adding molecular agents, in particular, hydrogen.
6. The construction of gas-discharge tubes and power supplies should be improved.

Hence, we may hope that the optimization of the exciting conditions that neutralize the influence of N_{e0} and N_{ms0} would increase the average generation power of CVL and its efficiency. It is worth noting that the possibility of obtaining high linear power $P_{las} > 100$ W m^{-1} in tubes of moderate diameter has been illustrated in [251].

10
Appendix D: Neutron Transmutation Doping of Silica

The method of neutron transmutation doping (NTD) of silica is based on nuclear transmutation of isotopes of semiconductor materials by capturing of slow (thermal) neutrons [265–268]. NTD is performed by irradiating samples or ingots of semiconductor crystals by neutron flux in nuclear reactors. By capturing neutron, isotope atom transmutes into other isotope with mass number greater by unity:

$$\Phi \sigma_i {}^A_Z N = {}^{A+1}_Z N \qquad (10.1)$$

Here, Φ [cm^{-2}] is the integral flux (doze) of thermal neutrons; σ_i [cm^2] is the cross-section of thermal neutron capture by given isotope; ${}^A_Z N$ and ${}^{A+1}_Z N$ [cm^{-3}] are the concentrations of initial and final products of the reaction, respectively; Z is the nuclear charge, and A is its mass number. If the resulting isotope ${}^{A+1}_Z N$ is stable, then such nuclear reaction does not lead to doping. The case of obtaining unstable isotope is most interesting. In a certain time lapse (the half-life period) it transmutes into nuclear of new element, whose number is greater by unity (${}^{A+1}_{Z+1} N$) in the case of decay or less by unity (${}^{A+1}_{Z-1} N$) if it is electron capture.

As an example, we may consider the reaction that results in phosphorus doping of silica:

$$^{30}_{14}\text{Si} + n = {}^{31}_{14}\text{Si} - \beta^- (2.62 \text{ h}) \rightarrow {}^{31}_{15}\text{P} \qquad (10.2)$$

The interest to NTD is caused by its two main advantages over the conventional metallurgical methods of doping. The first advantage is high accuracy of doping because the concentration of dopant at constant neutron flux is proportional to the duration of irradiation, which can be controlled with high accuracy. The second advantage is high uniformity of doping, which is determined by the random character of isotope distribution, a small cross-section of neutron capture and the uniformity of neutron flux. In view of the fact that σ_i are in the range of 10^{-23}–10^{-24} cm^{-2}, it is easy to find that at the maximal fluxes of thermal neutrons in modern nuclear reactors and reasonable duration of irradiation, the concentration of phosphorus dopant in silica would be of the order of 10^{15} cm^{-3}. It is sufficient

Laser Isotope Separation in Atomic Vapor. P. A. Bokhan, V. V. Buchanov, N. V. Fateev,
M. M. Kalugin, M. A. Kazaryan, A. M. Prokhorov, D. E. Zakrevskiĭ
Copyright © 2006 WILEY-VCH Verlag GmbH & Co. KGaA, Weinheim
ISBN: 3-527-40621-2

for many important practical applications, especially for producing high-voltage diodes and thyristors.

In the USA and Europe, the industrial technology has been created for producing hundreds ton of NTD-Si yearly in special scientific investigation reactors. In the USSR, the original technology for industrial production of NTD-Si was developed on the basis of nuclear high-power reactor RBMK-1000.

The NTD process, however, is not finished by irradiating samples or ingots in a nuclear reactor. The so-called fast neutrons are present in the reactor radiation spectrum, which possess high energy and lead to the formation of radiation damages or even "disordered domains." Annealing of these defects is a complicated technological problem because they produce complexes with dopants that are present in source material. It is necessary to develop different regimes of annealing (temperature, duration, and atmosphere) for different semiconductor materials or for the same material with different content of certain deep residual impurities (oxygen, carbon).

It was mentioned that the NTD method exhibits high accuracy of doping due to the linear dependence of doping concentration on radiation dose. In Fig. 10.1, the concentration of doped phosphorus (measured by the Hall effect) is shown for silica subjected to different radiation doses in a nuclear reactor.

One more important advantage of an NTD is the high uniformity of the dopant distribution. In ordinary metallurgical doping of semiconductors, the dopant is introduced into melt, which is followed by the growth of crystals. In this case, the difficulties of obtaining the uniform dopant distribution are of principal character. They are connected with the instability of the solid–melt interface in the doped crystals and with inevitable temperature gradient in the growing ingot that arise between its center and peripheral domains. These difficulties are well pronounced at greater diameter of ingots.

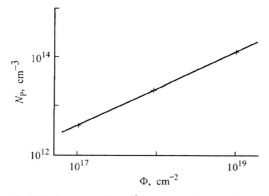

Fig. 10.1 Concentration of P atoms in Si crystals irradiated by various radiation doses of neutron flux followed by annealing at 800 °C for an hour (data from [269]).

The uniform doping in the NTD method is explained by the random distribution of isotopes over crystal lattice, the uniformity of neutron flux (the samples are rotated around their axis and simultaneously pulled through the active zone of reactor in the process of irradiation), and the small value of σ_i. The product $\sigma_i N_i$ summed over all stable isotopes of particular semiconductor defines the coefficient γ of linear absorption of neutrons, that is the value of the "block-effect" (screening of inner parts of ingot by outer parts). The parameter γ makes it possible to calculate the maximal dimension of crystal that can be doped by the NTD method at prescribed uniformity. For example, for Si we have $\gamma = 4.5 \times 10^{-3}$ cm^{-1}, which provides the macroscopic uniformity of doping not worse than 10 % even for large ingots with a diameter of 200 mm.

The NTD method also demonstrates the obvious advantage in the microscopic distribution of specific resistance even over the reference resistors manufactures by "metallurgical" method (Fig. 10.2). High macroscopic and microscopic uniformity of doping provided the wide employment of NTD-Si (including materials obtained in the reactor RBMK-1000) in industrial production of high-voltage converters.

In recent years, works on NTD of germanium with artificially changed isotopic composition were developed. The main idea is obtaining NTD-Ge with different polarities of conductivity and compensation factors, however, with uniform distribution of dopants. The possibilities of isotope engineering in silica are limited because NTD may only be performed with the ^{30}Si isotope, whose natural content is ≈ 3.1 %. Hence, the enrichment of natural Si by this isotope would result in 30-fold limiting concentration of phosphorus doped by the NTD method.

The prospects of works on NTD-Si and NTD-Ge are related to the creation of multilayer structures. For example, the possibility of creating alternating layers

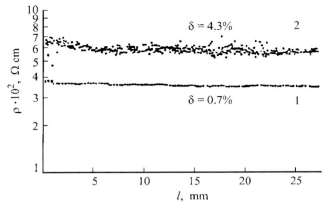

Fig. 10.2 Microscopic distribution of the specific resistance of reference resistors made from germanium by (1) NTD method and (2) "metallurgical" method (data from [270]).

from ^{70}Ge and ^{74}Ge isotopes in Si/Si$_{1-x}$Ge$_x$ heterojunctions or from the three germanium isotopes ^{70}Ge, ^{72}Ge, and ^{74}Ge in purely germanium structure followed by the NTD procedure. This would result in producing (p–i–n)-superlattices. The main feature of this method is separation of the processes of superlattice growing and doping. In conventional epitaxial methods these processes are combined, which results in mutual negative influence, namely, the smeared junction boundary due to dopant "emergence" in the growing process and the greater number of defects in the growing layer due to the influence of dopant comprised in it. In the method based on materials with controlled artificially changed isotopic composition, the layers grow without dopants, hence, they are expected to have perfect structure because different isotopes are chemically identical. The following NTD process and annealing occur at considerably low temperature, which would not smear the junction boundaries and worsen the structure of layers.

Obtaining NTD-Si in industrial quantities and employment of NTD-Si and NTD-Ge in producing high-power thyristors, detectors of nuclear particles and infrared radiation, deeply cooled thermistors and bolometers are actual problems.

11
Appendix E: Employment of Boron Isotopes in Microelectronics

Boron is widely used in microelectronic technology for producing p-type domains in silica. There are two types of boron: amorphous boron is brown powder and crystal boron looks like white crystals. When introduced to silica (or germanium), boron forms acceptor dopant with a high limiting solubility. Elementary boron metal cannot be introduced to silica (for example, by diffusion method in gas flux) because of a low vapor pressure, hence, its intermediate chemical compounds are conventionally used, such as B_2O_3 (boron anhydride), BF_3, and other boron haloids.

In Fig. 11.1, the limiting solubility versus temperature in silica is shown for various elements. Boron possesses high limiting solubility and its surface concentration may reach $N_B = 4 \times 10^{20}$ cm^{-3}. The radius of tetrahedral boron in silica is 0.88×10^{-4} µm and the corresponding radius for silica is 1.17×10^{-4} µm. Hence, the presence of a large number of boron atoms leads to the formation of defects induced by mechanical stresses, which at a sufficiently high volume or surface density substantially damage the monocrystal silica lattice. They may also form macroscopic defects (dislocations) with dimensions far greater than the width of the layer doped with the boron atoms. The excessive concentrations of boron impurity stimulate noticeable damages in the initial regular structure of silica monocrystal. In order to avoid this, the boron concentration should not be greater than 5×10^{19} cm^{-3}.

It is also necessary to take into account that the initial purity of the source of boron (that is, the absence of impurities of other chemical elements) should be quite high. Even a small quantity of additional impurities in boron alloy may result in a substantial uncontrolled contamination of silica single crystal.

In transferring to monocrystal plates on the basis of mono-isotopic silica, its lattice formed by atoms (ions) of the same dimension becomes highly ordered and the density of point defects becomes low. We may expect that in such a perfect lattice it would be possible to provide conditions preventing the origin of large-scale volume dislocations (presently, it is the high density of dislocations that limits the commercial production of silica plates).

Laser Isotope Separation in Atomic Vapor. P. A. Bokhan, V. V. Buchanov, N. V. Fateev,
M. M. Kalugin, M. A. Kazaryan, A. M. Prokhorov, D. E. Zakrevskiĭ
Copyright © 2006 WILEY-VCH Verlag GmbH & Co. KGaA, Weinheim
ISBN: 3-527-40621-2

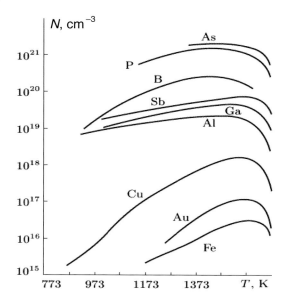

Fig. 11.1 The limiting solubility of additives in silica.

The requirements for the maximal possible degree of perfection of the produced mono-isotopic lattice are completely applicable to subsequent technological operations with silica plates. In particular, in doping acceptor material it is reasonable to use isotopically pure boron ^{10}B or ^{11}B free of foreign atoms; limiting its concentration in silica at the optimal level. This provides minimal distortions of the crystal structure of silica matrix, minimizes the thickness of layer with boron acceptor (and the total thickness of (p–n)-junction), increases the heat conductivity of the acceptor layer, hence, enhances the efficiency of heat removal to the plate surface, and, finally, increases the transistor switching power.

12
Appendix F: Employment of Boron in Nuclear Fuel Cycle Equipment

Natural boron includes ^{10}B isotopes, whose nuclei have a large cross-section of neutron absorption. Therefore, boron is often used in equipment for regulating and controlling industrial nuclear reactors as a burnable absorber.

It is known that the heat energy released in a nuclear reactor of the atomic power plant is determined by the intensity of the chain reaction that forms the neutron flux with prescribed density. The state of a chain reaction is characterized by the special multiplication constant k.

At $k = 1$ (the critical state of reactor), the number of neutrons disappearing in unit time (captured by nuclei of fissile material) is equal to the number of neutrons that happen to grow in nuclear fission. In this case, the thermal energy of the reactor remains constant. At $k > 1$ (the above-critical state), multiplication of neutrons is observed, the density of neutron flux grows exponentially, and the power of reactor rises. The regime with $k < 1$ (the subcritical state) is characterized by an exponential reduction of the density of neutron flux, which is accompanied with a reduction of thermal power.

If there are technical equipment and methods for the varying factor k with the prescribed accuracy, then the possibility of controlling the reactor arises. It is possible to obtain the required power by varying k. Then at the new level of power, the value $k = 1$ is restored, thus stabilizing the heat release.

A controlled reactor response to the variation of k is a very complicated function of a number of interrelating parameters, which is described by the parameter ρ, which is called reactivity. In general form, the dependence $\rho(k)$ is defined via the so-called excessive multiplication parameter σk, which is the deviation of k from the steady value $k = 1$ corresponding to the critical reactor state. The temporal evolution of neutron density after the step variation of k from 1 to $1 + \sigma k$ is described by the exponential equation with the exponential σk.

At $\sigma k > 0$, density of neutrons exponentially grows. At $\sigma k < 0$, it reduces exponentially, and at $\sigma k = 0$, the neutron density becomes constant.

The value and sign of σk in the active reactor zone are adjusted by a control system. The creation of highly efficient control, regulation, and protection systems

Laser Isotope Separation in Atomic Vapor. P. A. Bokhan, V. V. Buchanov, N. V. Fateev,
M. M. Kalugin, M. A. Kazaryan, A. M. Prokhorov, D. E. Zakrevskiĭ
Copyright © 2006 WILEY-VCH Verlag GmbH & Co. KGaA, Weinheim
ISBN: 3-527-40621-2

(CRPS) is the main measure providing safe, reliable, and efficient operation of nuclear reactors in high power atomic power plants. The active zone of operating reactor contains an excessive quantity of fissile material with respect to the critical mass, hence, the main task of CRPS is to provide guaranteed nuclear safety. For solving this problem in all conditions and situations, CRPS continuously compensates an excess of fission material, the reduction of the initially loaded fuel (usually UO_2) caused by its gradual "burning" in a fission process, accumulation of nuclear debris accompanied by reactor "poisoning," the influence of temperature effects leading to variations in the temperature coefficients of reactivity, and so forth.

The reactor differs from other regulated objects in that its operating power may vary in very wide limits (for example, in the subcritical state, the minimal power of stopped reactor may amount to 10^{-11}–10^{-10} of its nominal power). Hence, the main dynamic response of reactor as a control target is the parameter "reactivity-power."

The level of requirements for the regulating system is very high. For different states and operation modes (subcritical, reactor is stopped; reactor start-up; reactor riding up; operation in the nominal regime; reactor terminated), particular complexes of controlled and regulated parameters are used. Requirements for the control and protection system in the mentioned regimes are also different. In power operation regimes (in particular, close to nominal operation), such parameters as the reactor power and its distribution over the volume of active zone; the temperature of fuel elements, coolant, and the construction elements of reactor; the pressure and consumption of coolant should be maintained with a required accuracy. In this case, the main requirements for the regulation system are: high accuracy of maintaining the steady-state regimes; low tolerance of regulation (because at the nominal power many parameters are close to the maximum permissible values) in the course of regulation, deviations of the transient processes and the set-up time of a chosen regime should be minimal; high reliability of the regulation system should be provided.

In high-power reactors (for example, RBMK-1000), regulation is realized through changing the multiplication factor k by varying the rate of neutron absorption. It is done by changing the number of absorbers (moving them in or out) that reside in the active zone and contain chemical elements, whose nuclei have large cross-section of neutron absorption (without producing secondary neutrons).

The absorber, after capturing neutrons, decays producing lighter elements, that is, "burns out." The burning absorber is usually accommodated in the regulating and control rods. The life time of the burning absorber should agree with nuclear fuel burnup.

In industrial nuclear reactors, boron and its compounds are used as a solid absorbing material for regulating rods. One can see from Table 12.1 that neutrons are mainly absorbed by the ^{10}B isotope, whose relative content in natural boron is

Table 12.1 Absorbing materials and their properties

Material	Melting temperature (°C)	Density (g cm^{-3})	Microscopic absorption cross-section of thermal neutrons (b)
Boron (^{10}B)	2300	2.4	3840
Boron (natural mixture)	2300	2.4	755
Cadmium	321	8.6	2450
Cobalt	1495	8.7	37
Dysprosium	1400	8.6	950
Erbium	1550	9.1	173
Europium	900	5.2	4300
Gadolinium	1350	8.0	46000
Hafnium	2220	13.1	105
Holmium	1500	8.8	65
Indium	156	7.3	196
Iridium	2442	22.4	440
Lithium	186	0.5	71
Osmium	3000	22.5	15.3
Rhenium	3180	21.0	864
Rhodium	1960	12.4	156
Samarium	1052	7.8	5600
Argentum	961	10.5	63
Tantalum	2996	16.6	21
Thulium	1650	9.4	127
Tungsten	3410	19.3	19
Zircaloy 2[1)]	1852	6.6	0.180
Iron	1535	7.9	2.53

[1)] Is presented for a comparison with absorbing materials
(zircaloy 2 comprises 98 % Zr; 1.5 % Sn; 0.15 % Fe; 0.10 % Cr; 0.05 % Ni)

about 18 %. Pure amorphous or crystal boron is used as well as boron comprised in such compounds as $Na_2B_4O_7$, H_3BO_3, B_4C, BN, and metal compounds. For enhancing the absorption efficiency, natural boron may be enriched to 90 % and

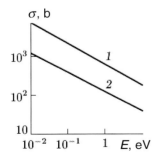

Fig. 12.1 The absorption cross-section of (1) ^{10}B isotope and (2) natural boron versus the energy of neutrons.

greater content of ^{10}B isotope. The cross-section of neutron absorption by boron versus neutron energy is shown in Fig. 12.1. Boron carbide pressed in tablets or briquettes with the following high-temperature agglomeration (at $T > 2100\ °C$) is used as a absorption material in producing control rods. 1 cm^3 of natural boron with the density of 2.5 g cm^{-3} comprises 2.3×10^{22} nuclei of ^{10}B.

If neutron is absorbed by boron nucleus, then the reaction occurs, in which helium and lithium nuclei arise: ^{10}B + ^{1}n $\rightarrow$ ^{4}He + ^{7}Li. Hence, the irradiation of control rods by neutron flux is accompanied with helium released from the boron carbide powder or tablets. The yield of helium relative to the total number of helium nuclei produced in the considered reaction is determined by various factors, including the degree of boron carbide enrichment by the ^{10}B isotope, the degree of burning (see the scheme of reaction), spectral distribution of neutrons over energy, and so forth.

References

1 *Isotopes: Properties, Production and Applications*, Ed. V. Yu. Baranov, IzdAT, Moscow, **2000** [in Russian]
2 S. Mrozowski, *Z. für Phys.* 786 (**1932**), p. 826
3 K. Zuber, *Nature* 136 (**1935**), p. 796
4 C. C. McDonald, H. E. Gunning, *J. Chem. Physics* 20 (**1952**), p. 817
5 V. S. Letokhov, R. V. Ambartzumian, *IEEE J. Quant. Electr.* QE-7 (**1971**), p. 305
6 R. V. Ambartzumian, V. S. Letokhov, *Appl. Opt.* 11 (**1972**), p. 354
7 R. V. Ambartzumian, V. P. Kalinin, V. S. Letokhov, *JETP Lett.* 13 (**1971**), p. 217
8 V. S. Letokhov, *Science* 180 (**1973**), p. 451
9 V. S. Letokhov, *Opt. Commun.* 7 (**1973**), p. 59
10 N. V. Karlov, A. M. Prokhorov, *Sov. Phys. Usp.* 19 (**1976**), p. 285
11 V. S. Letokhov, C. B. Moore, *Sov. J. Quant. Electron.* 6 (**1976**), p. 129
12 V. S. Letokhov, C. B. Moore, *Sov. J. Quant. Electron.* 6 (**1976**), p. 259
13 N. G. Basov, E. M. Belenov, V. A. Isakov et al., *Sov. Phys. Usp.* 20 (**1977**), p. 209
14 V. S. Letokhov, V. A. Michin, A. A. Puretzky, *Plasma Chemistry*, AtomEnergoizdat, Moscow, Vol. 4, **1977**, p. 3 [in Russian]
15 N. V. Karlov, B. B. Krynetskii, V. A. Michin et al., *Sov. Phys. Usp.* 28 (**1976**), p. 385
16 V. S. Letokhov, *Sov. Phys. Usp.* 26 (**1978**), p. 23
17 N. V. Karlov, *Proceedings of Lebedev Physical Institute* 114 (**1979**), p. 3 [in Russian]
18 V. S. Letyokhov, *Non-Linear Selective Photoprocesses in Atoms and Molecules*, Nauka, Moscow, **1983** [in Russian]
19 S. I. Yakovlenko, *Quant. Electron.* 28 (**1998**), p. 945
20 V. A. Mishin, in *Isotopes: Properties, Production and Applications*, Ed. V. Yu. Baranov IzdAT, Moscow, **2000**, p. 308 [in Russian]
21 J. Devis, R. Devis, in *Laser Program Annual Report-1976, UCRL-50021-76*, Ed. P. E. Coyle, Lawrence Livermore National Laboratory, CA, 8/3-8, **1977**
22 M. Spaeth, in *Laser Program Annual Report-1976, UCRL-50021-76*, Ed. P. E. Coyle, Lawrence Livermore National Laboratory, CA, 8/3-8, **1977**
23 S. Shirayama, H. Ueda, T. Mikatsura et al., *Proc. SPIE* 1225 (**1990**), p. 279
24 N. Morioka, *Proc. SPIE* 1859 (**1993**), p. 2
25 N. Camarcat, A. Lafon, J. Perves et al., *Proc. SPIE* 1859 (**1993**), p. 14
26 C. A. Haynam, B. J. Comaskey, J. Conway et al., *Proc. SPIE* 1859 (**1993**), p. 24
27 D. Levron, A. Bar-Shalom, Z. Burshtein et al., *Proc. SPIE* 1859 (**1993**), p. 69
28 G. Forest, *Laser Focus* 22 (**1986**), p. 23
29 V. Kiernan, *Laser Focus World* 33 (**1977**), p. 78
30 B. Grant, *Photonics Spectra* 31 (**1997**), p. 46
31 A. A. Isaev, M. A. Kazaryan, G. G. Petrash, *JETP Lett.* 16 (**1972**), p. 29
32 V. M. Batenin, V. V. Buchanov, M. A. Kazaryan et al., *Lasers on Self-Terminating Transitions of Metal Atoms* Nauchnaya kniga, Moscow (**1988**) [in Russian]
33 C. E. Little, *Metal Vapour Lasers*, Wiley, New York, **1999**
34 A. N. Soldatov, V. I. Solomonov, *Self-Terminating Gas-Discharge Metal Vapor Lasers*, Nauka, Novosibirsk, **1985** [in Russian]
35 *Proc. SPIE Int. Soc. Opt. Eng.* 1859 (**1993**)

Laser Isotope Separation in Atomic Vapor. P. A. Bokhan, V. V. Buchanov, N. V. Fateev, M. M. Kalugin, M. A. Kazaryan, A. M. Prokhorov, D. E. Zakrevskiĭ
Copyright © 2006 WILEY-VCH Verlag GmbH & Co. KGaA, Weinheim
ISBN: 3-527-40621-2

36 N. A. Lyabin, A. D. Chursin, S. A. Ugolnikov et al., *Quantum Electron.* 31 **(2001)**, p. 191

37 R. E. Grove, *Copper vapour laser overview, Laser program annual report-1979* LLNL: Livermore, CA, **1980**, pp. 9-4 to 9-5

38 J. L. Devis, B. W. Shore, *Technical and systems highlights, Laser program annual report-1980, UCRL-5002180* LLNL: Livermore, CA, **1981**, pp. 10-13 to 10-17

39 B. E. Warner, *Status of copper vapour technology at Lawrence Livermore National Laboratory, CLEO 1991* Tech. Dig., Opt. Soc. Am., Wasington, DC, **1991**, pp. 516–518

40 A. Kearsley, *Proc. SPIE* 1225 **(1990)**, p. 270

41 N. V. Sabotinov, N. K. Vuchkov, D. N. Asadjov, *Proc. SPIE* 1225 **(1990)**, p. 289

42 I. L. Blass, R. E. Bonanno, R. P. Hackel, P. R. Hammond, *Appl. Opt.* 31 **(1992)**, p. 6993

43 D. Doizi, *Pulsed Metal Vapour Lasers*, Kluwer, Dordrecht, **1996**, p. 303

44 S. V. Vasil'ev, V. A. Mishin, T. V. Shavrova, *Quantum Electron.* 27 **(1997)**, p. 126

45 Y. Tabuta, K. Hara, S. Ueguri, *Proc. SPIE* 1628 **(1992)**, p. 32

46 P. A. Bokhan, V. V. Buchanov, D. E. Zakrevsky et al., *JETP Lett.* 71 **(2000)**, p. 483

47 P. A. Bokhan, D. E. Zakrevsky, N. V. Fateev et al., *Proceedings of the 5th International Scientific Conference Physical and Chemical Processes on Selection of Atoms and Molecules*, TsNIIatominform, Moscow, Zvenigorod, **(2000)**, pp. 98–101

48 N. A. Lyabin, *Atmos. Oceanic Opt.* 13 **(2000)**, p. 258

49 O. P. Maximov, V. A. Mishin, A. G. Mashkunov et al., *Technical Digest of International Conference Lasers-97*, New Orleans, Lousiana, **1997**

50 P. A. Bokhan, D. E. Zakrevsky, S. A. Kochubei et al., *Quantum Electron.* 31 **(2001)**, p. 132

51 A. I. Moshkunov, D. E. Zakrevsky, G. G. Rahimov et al., *Proceedings of the 5th International Scientific Conference Physical and Chemical Processes on Selection of Atoms and Molecules*, TsNIIatominform, Moscow, Zvenigorod, **2000**, pp. 94–97

52 S. G. Anderson, *Laser Focus World* 37 (1) **(2001)**, p. 88

53 S. M. Kobtsev, A. V. Korablev, S. V. Kukarin. et al., *Proc. SPIE* 4353 **(2001)**, p. 189

54 P. A. Bokhan, D. E. Zakrevsky, A. Yu. Stepanov et al., *Proceedings of the 6th International Scientific Conference Physical and Chemical Processes on Selection of Atoms and Molecules*, TsNIIatominform, Moscow, Zvenigorod, **2001**, pp. 101–104

55 P. A. Bokhan, D. E. Zakrevsky, V. A. Kim et al., *Proceedings of the 6th International Scientific Conference Physical and Chemical Processes on Selection of Atoms and Molecules*, TsNIIatominform, Moscow, Zvenigorod, **2001**, pp. 105–107

56 P. A. Bokhan, V. V. Buchanov, D. E. Zakrevsky et al., *Proceedings of the 4th International Scientific Conference Physical and Chemical Processes on Selection of Atoms and Molecules*, TsNIIatominform, Moscow, Zvenigorod, **1999**, pp. 77–80

57 N. S. Demidova, V. A. Mishin, *Sov. Tech. Phys. Lett.* 23 **(1977)**, p. 42 [in Russian]

58 S. K. Borisov, M. A. Kuz'mina, V. A. Mishin, *Proceedings of the 4th International Scientific Conference Physical and Chemical Processes on Selection of Atoms and Molecules*, TsNIIatominform, Moscow, Zvenigorod, **1998**, p. 40

59 V. V. Buchanov, M. A. Kazaryan, M. M. Kalugin et al., *Laser Physics* 11 **(2001)**, p. 1332

60 V. V. Buchanov, M. A. Kazaryan, M. M. Kalugin et al., *Proceedings of the 6th International Scientific Conference Physical and Chemical Processes on Selection of Atoms and Molecules*, TsNIIatominform, Moscow, Zvenigorod, **2001**, p. 108

61 V. V. Buchanov, M. A. Kazaryan, M. M. Kalugin. et al., *Technical Digest of International Conference on Lasers 2001* Tucson, AZ, **2001**, p. 34

62 S. K. Borisov, M. A. Kuz'mina, V. A. Mishin, *Quantum Electron.* 25 **(1995)**, p. 695

63 S. K. Borisov, V. A. Mishin, *Proceedings of Institute of General Physics* 24 **(1990)**, p. 3 [in Russian]

64 J. A. Paisner, *Appl. Phys. B* 46 **(1988)**, p. 253

65 P. T. Greenland, *Contemp. Phys.* 31 **(1990)**, p. 405

66 J. L. Emmett, W. F. Krupke, J. I. Davis, *IEEE J. Quantum Electron.* QE-20 **(1984)**, p. 891

67 J. I. Davis, E. B. Rockower, *IEEE J. Quantum Electron.* QE-18 (**1982**), p. 233
68 A. N. Tkachev, S. I. Yakovlenko, *Quantum Electron.* 32 (**2002**), p. 614
69 N. S. Demidova, V. A. Mishin, *Proceedings of the 3th International Scientific Conference Physical and Chemical Processes on Selection of Atoms and Molecules*, TsNIIatominform, Moscow, Zvenigorod, **1998**, p. 59
70 B. M. Smirnov, *Excited Atoms*, Energoizdat, Moscow, **1982** [in Russian]
71 A. F. Bernhardt, *Appl. Phys.* 9 (**1976**), p. 19
72 W. A. Wijngaarden, J. Li, *Rhys. Rev. A* 49 (**1994**), p. 1158
73 F. Kh. Gel'mukhanov, A. M. Shalagin, *JETP Lett.* 29 (**1979**), p. 711
74 V. D. Antsygin, S. N. Atutov, F. Kh. Gel'mukhanov et al., *JETP Lett.* 30 (**1979**), p. 243
75 S. N. Atutov, P. I. Chapovsky, A. M. Shalagin, *Opt. Commun.* 43 (**1982**), p. 265
76 V. Yu. Baranov, E. P. Velikhov, A. M. Dykhne et al., *JETP Lett.* 31 (**1980**), p. 445
77 V. N. Panfilov, V. P. Strunin, P. L. Chapovskii, *JETP* 58 (**1983**), p. 510
78 V. S. Letokhov, *Isotopes: Properties, Production, Applications*, Ed. V. Yu. Baranov, IzdAT, Moscow, **2000**, p. 291 [in Russian]
79 W. T. Walter, M. Piltch, N. Solimene et al., *Bull. Am. Phys. Soc.* 11 (**1966**), p. 113
80 R. S. Anderson, L. W. Springer, T. W. Karras, *IEEE J. Quantum Electron.* QE-11, 56D (**1975**)
81 C. Konagai, H. Kimura, N. Aoki et al., *Y. Proc. 15th Ann. Meeting Laser Soc.* Laser Soc. Jpn, Jpn. Osaka, 15 **1995**, p. 112
82 C. Konagai, N. Aoki, Y. Sano, *Pulsed Metal Vapour lasers*, Kluwer, Dordrecht, **1996**, p. 371
83 J. J. Chang, B. E. Warner, C. D. Boley, E. P. Dragon, in *Pulsed Metal Vapour Lasers*, Kluwer, Dordrecht, **1996**, p. 101
84 R. P. Hackel, B. E. Warner, *Proc. SPIE Int. Soc. Opt. Eng.* 1859 (**1993**), p. 120
85 D. R. Jones, A. Maitland, C. E. Little, *IEEE J. Quantum Electron.* QE-30 (**1994**), p. 2385
86 J. J. Chang, *Appl. Opt.* 32 (**1993**), p. 5230
87 N. V. Kravtsov, *Quantum Electron.* 31 (**2001**), p. 661
88 K. Fukuta, S. Fujikawa, K. Yasui, *Technical Digest Conference on Advanced Solid-State Lasers*, Seattle, USA, **2001**, p. 12
89 M. Sato, S. Natio, H. Machida et al., *TOPS* 26 (**1999**), p. 2
90 Y. Axiyama, T. Takase, H. Ynasa, *Technical Digest CLEO 99*, San Francisco, USA, **1999**, p. 31
91 S. P. Velsko, C. A. Ebbers, B. Comaskey et al., *Appl. Phys. Lett.* 64 (**1994**), p. 3086
92 J. J. Chang, E. P. Dragon, C. A. Ebbers et al., *An efficient diode-pumped YAG:Nd laser with 451 W of CW IR and 182 W of pulsed green output, UCRL-JC-129704*, Preprint LLNL **1998**
93 J. J. Chang, E. P. Dragon, I. L. Bass, *Technical Digest CLEO 98*, Washington, DC, **1998**, p. CPD2-2
94 U. Brackmann, *Lambdachrome. Laser Dyes* Lambda Physik, **1996**
95 *High-Power Dye Lasers (Springer Series in Optical Sciences, v. 65)*, Ed. F. J. Duarte, Springer, Berlin, Heidelberg, **1991**
96 G. Erbert, I. Bass, R. Hackel et al., *J. Techn. Dig. CLEO 91* Washington, DC, **1991**, p. 390
97 W. J. Wadsworth, D. W. Coutts, C. E. Webb, *Appl. Opt.* 38 (**1999**), p. 6904
98 K. Scheibner, C. Haynam, E. Worden et al., *Laser Isotope Purification of Lead for Use in Semiconductor Chip Interconnects, UCRL–JC–122657* Preprint LLNL **1996**
99 P. A. Bokhan, *Proceedings of the 4th International Scientific Conference Physical and Chemical Processes on Selection of Atoms and Molecules*, TsNIIatominform, Moscow, Zvenigorod, **1999**, pp. 106–111
100 G. L. Khorasanov, A. P. Ivanov, A. I. Blokhin et al., *Proceedings of the 4th International Scientific Conference Physical and Chemical Processes on Selection of Atoms and Molecules*, TsNIIatominform, Moscow, Zvenigorod, **1999**, pp. 239–243
101 V. I. Derzhiev, V. M. Dyakin, R. I. Ilkaev, *Quantum Electron.* 32 (**2002**), p. 619
102 I. S. Grigoriev, E. B. Gelman, A. F. Semerok et al., *Proceedings of the 5th International Scientific Conference Physical and Chemical Processes on Selection of Atoms and Molecules*, TsNIIatominform, Moscow, Zvenigorod, **2000**, p. 88
103 A. B. Dyachkov, S. K. Kovalevich, A. V. Pesnya et al., *Proceedings of the 5th International Scientific Conference Physical and Chemical Processes on Selection of Atoms and Molecules*, TsNIIatominform, Moscow, Zvenigorod, **2000**, pp. 112–114

104 M. Compte, J. De Lamare, A. Petit et al., Proc. SPIE Int. Soc. Opt. Eng. 1859 (1993), p. 79
105 I. S. Grigoriev, Light Isotope Separation. Preprint 6246/12 Kurchatov Inst. At. Energy, Moscow, 2002
106 H. Bruesselbach, D. S. Sumida, M. Mangir, Technical Digest CLEO 97, Washington, DC, 1997, p. CPD34-2
107 S. A. Kostrisa, V. A. Michin, Quantum Electron. 25 (1995), p. 516
108 V. I. Derzhiev, V. A. Kuznetsov, L. A. Mikhal'tsov et al., Quantum Electron. 26 (1996), p. 751
109 V. I. Derzhiev, V. M. Dyakin, R. I. Ilkaev et al., Quantum Electron. 33 (2003), p. 553
110 I. S. Grigoriv, A. B. D'yachkov, V. P. Labozin et al., Proceedings of the 7th International Scientific Conference Physical and Chemical Processes on Selection of Atoms and Molecules, TsNIIatominform, Moscow, Zvenigorod, 2002, pp. 177–184
111 H. Okabe, Photochemistry of Small Molecules, Wiley, New York, 1978
112 H. S. W. Massey, E. H. S. Burhop, Electronic and Ionic Impact Phenomena, Clarendon, Oxford, 1952
113 D. R. Herschbach, Adv. Chem. Phys. 10 (1966), pp. 319–393
114 E. Bauer, F. R. Fischer, F. R. Gilmore, J. Chem. Phys. 51 (1969), p. 4173
115 E. A. Gislason, J. G. Sachs, J. Chem. Phys. 62 (1975), p. 2678
116 A. A. Zembekov, E. E. Nikitin, U. Havemann, L. Zulike, in Chemistry of Plasma, Ed. B. M. Smirnov, AtomEnergoizdat, Moscow, 6,), p. 3 [in Russian]
117 T. Laurent, P. D. Naik, H.-R. Volpp et al., Chem. Phy. Lett. 236 (1995), p. 343
118 X. Huang, J. Zhao, G. Xing et al., J. Chem. Phys. 104 (1996), p. 1338
119 D. K. Liu, K. C. Lin, J. Chem. Phys. 107 (1997), p. 4244
120 M. Motzkus, G. Pichler, K. L. Kompa et al., P. J. Chem. Phys. 106 (1997), p. 9057
121 L. H. Fan, J. J. Chen, Y. Y. Lin, W. T. Luh, J. Phys. Chem. A 103 (1999), p. 1300
122 P. A. Bokhan, D. E. Zakrevsky, N. V. Fateev, JETP Lett. 75 (2002), p. 170
123 W. H. Breckenridge, J. H. Wang, Chem. Phys. Lett. 137 (1987), p. 195
124 D. K. Liu, K. C. Lin, Chem. Phys Lett. 304 (1999), p. 336

125 O. Nedeles, M. Giroud, Chem. Phys. Lett. 165 (1990), p. 329
126 N. Bras, J. Butaux, J. C. Jeannet, D. Perrin, J. Chem. Phys. 85 (1) (1986), p. 280
127 W. H. Breckenridge, A. M. Renlund, J. Phys. Chem. 83 (1979), p. 1145
128 W. H. Breckenridge, J. H. Wang, J. Chem. Phys. 87 (5) (1987), p. 2630
129 W. H. Breckenridge, A. M. Renlund, J. Phys. Chem. 82 (1978), p. 1474
130 W. H. Breckenridge, T. W. Broadbent, D. S. Moore, J. Phys. Chem. 79 (1975), p. 1233
131 S. Yamamoto, N. Nishimura, Bull. Chem. Soc. Jpn. 55 (1982), p. 1395
132 H. Umemoto, S. Tsunashima, H. Ikeda et al., J. Chem. Phys. 101 (6) (1994), p. 4803
133 H. Umemoto, K. Matsumoto, Chem. Phys. Lett. 236 (1995), p. 408
134 W. H. Breckenridge, A. M. Renlund, J. Phys. Chem. 83 (1979), p. 303
135 M. Gzajkowich, E. Walentynowich, L. Krause, J. Quant. Spectrosc. Radiat. Transfer 29 (1983), p. 113
136 H. Umemoto, N. Ohsako, Chem. Phys. 221 (1997), p. 209
137 W. H. Breckenridge, H. Umemoto, J. Phys. Chem. 87 (1983), p. 476
138 K. Kuwahara, H. Ikeda, H. Umemoto et al., J. Chem. Phys. 99 (4) (1993), p. 2715
139 M. Garay, J. M. Orea, A. G. Urena, Chem. Phys. 207 (1996), p. 451
140 D. Husain, J. Geng, F. Castanon et al., J. Photochem. Photobiol. A: Chem. 133 (2000), p. 1
141 S. A. Mitchel, P. A. Hacked, D. M. Rayner, J. Chem. Phys 86 (1987), p. 6853
142 W. Fang, Chem. Phys. Lett. 260 (1996), p. 565
143 K. Chen, C. S. Sung, J. Chang et al., Chem. Phys. Lett. 240 (1995), p. 17
144 V. N. Kondratiev, Rate Constants of Gaseous Reactions, Nauka, Moscow, 1970 [in Russian]
145 K. Sato, N. Ishida, T. Kurakata et al., Chem. Phys. 237 (1998), p. 195
146 D. Husain, A. X. Ioannou, M. Kabir, J. Photochem. Photobiol. A: Chem. 110 (1997), p. 213
147 T. G. Aardema, N. A. Asten, J. P. J. Driessen et al., Chem. Phys. 202 (1996), p. 377
148 M. L. Campbell, R. E. McClean, J.S.S. Harter, Chem. Phys. Lett. 235 (1995), p. 497

149 R. Matsui, K. Senba, K. Honma, *Chem. Phys. Lett.* 250 (**1996**), p. 560

150 T. F. Gallagher, G. A. Ruff, K. A. Safinua, *Phys. Rev. A* 22 (**1980**), p. 843

151 P. A. Bokhan, D. E. Zakrevsky, N. V. Fateev, *JETP* 98 (**2004**), p. 24

152 J. B. Boffard, M. D. Stewart, C. C. Lin, *Phys. Rev. A* 65 (**2002**), p. 062701

153 Eds. R. F. Stebbingts, F. B. Dunning, *Rydberg States of Atoms and Molecules*, Cambridge University Press, Cambridge, **1983**

154 M. A. Mazing, P. D. Serapinas, *Sov. Phys. JETP* 33 (**1971**), p. 294

155 D. C. Thompson, E. Weinberger, G. X. Xu et al., *Phys. Rev. A* 35 (**1987**), p. 690

156 D. C. Thompson, E. Kammermayer, B. P. Stoicheff et al., *Phys. Rev. A* 36 (**1987**), p. 2134

157 R. Kachru, T. W. Mossberg, S. R. Hartmann, *Phys. Rev. A* 21 (**1980**), p. 1124

158 T. F. Gallagher, R. E. Olson, W. E. Cooke et al., *Phys. Rev. A* 16 (**1977**), p. 441

159 A. N. Klucharev, A. V. Lazarenko, V. Vujnovic, *J. Phys. B: Atom. Mol. Phys.* 13 (**1980**), p. 1143

160 J. Boulmer, G. Baran, F. Devos et al., *Phys. Rev. Lett.* 44 (**1980**), p. 1122

161 B. M. Smirnov, *Introduction to Plasma Physics*, Nauka, Moscow, **1982** [in Russian]

162 A. I. Brodsky, *Isotope Chemistry*, AN SSSR, Moscow, **1957** [in Russian]

163 Y. Kuga, K. Takeuchi, *J. Chem. Phys.* 108 (**1998**), p. 4591

164 M. F. Bertino, J. P. Toennies, *J. Chem. Phys.* 110 (**1999**), p. 9186

165 R. A. Ogg, *J. Chem. Phys.* 15 (**1947**), p. 613

166 A. A. Viggiano, R. A. Morris, *J. Chem. Phys.* 100 (**1994**), p. 2748

167 H. J. Kim, Y. D. Park, W. M. Lee, *Plasma Chem. Plasma Process.* 20 (**2000**), p. 259

168 S. M. Anderson, F. S. Klein, F. Kaufman, *J. Chem..Phys.* 83 (**1985**), p. 1648

169 J. M. Barthez, A. V. Filikov, L. B. Frederriksen et al., *Can. J. Chem.* 76 (**1998**), p. 726

170 V. N. Kondratiev, E. E. Nikitin, *Kinetics and Mechanisms of Gas-Phase Reactions*, Nauka, Moscow, **1975** [in Russian]

171 *Physical Chemistry of Fast Reactions*, Ed. Levitt, Plenum, London, **1973**

172 J. Sommar, M. Hallquist, E. Ljungstrom, *Chem. Phys. Lett.* 257 (**1996**), p. 434

173 I. Szilagui, S. Dobe, T. Beses, *React. Kinet. Catal. Lett.* 70 (**2000**), p. 319

174 P. Fleurat-Lessard, J. C. Rayes, A. Bergeat et al., *Chem. Phys.* 279 (**2002**), p. 87

175 W. Hack, R. Jordan, *Chem. Phys. Lett.* 306 (**1999**), p. 111

176 Y. U. Viazovetsky, *Isotopes: Properties, Production and Applications*, Ed. V.Yu. Baranov, IzdAT, Moscow, **2000**

177 Yu. N. Molin, V. N. Panfilov, A. K. Petrov, *Infrared Photochemistry*, Nauka, Novosibirsk, **1985**

178 H. E. Gunning, *Can. J. Chem.* 36 (**1958**), p. 89

179 Y. U. Viazovetsky, S. A. Senchenkov, *Tech. Phys.* 68 (**1987**), p. 1643 [in Russian]

180 P. Cambpell, J. Billowes, I. S. Grant, *J. Phys. B: At Mol. Opt. Phys.* 30 (**1997**), p. 2351

181 Ya. I. Khanin, *Quantum Radiophysics: Dynamics of Quantum Generators* Sovetskoe Radio, Moscow, **1975** [in Russian]

182 V. S. Letokhov, V. P. Chebotaev, *Nonlinear Laser Spectroscopy*, Springer, Berlin, **1997**

183 A. N. Klucharev, M. L. Yanson, *Elementary Processes in Alkali Metal Plasma*, Energoizdat, Moscow, **1988** [in Russian]

184 C. C. Dobson, C. C. Sung, *Phys. Rev. A* 59 (**1999**), p. 3402

185 L. S. Vasilenko, V. P. Chebotaev, A. V. Shishaev, *JETP Lett.* 12 (**1970**), p. 113

186 L. Allen, J. H. Eberly, *Optical Resonance and Two Level Atoms*, Wiley, New York, **1975**

187 E. B. Salomon, *Spectrochim. Acta* 45B (1/2) (**1990**), p. 37

188 *Handbook of Physics*, Eds. I. S. Grigoriev, E. Z. Meylikhov, Energoizdat, Moscow, **1991** [in Russian]

189 L. D. Landau, E. M. Lifshitz, *Quantum Mechanics*, Pergamon, Oxford, **1977**

190 I. I. Sobelman, *Introduction to the Theory of Atomic Spectra*, Pergamon, Oxford, P**1973**

191 E. T. Getty, *Appl. Phys. Letters* 7 (**1965**), p. 6

192 P. L. Kelley, H. Kildal, H. R. Scholssberg, *Chem. Phys. Lett.* 27 (**1974**), p. 62

193 K. Shimoda, *Appl. Phys.* 9 (**1976**), p. 239

194 E. M. Dianov, *Vestnik RAN* 71 (**2001**), p. 1030 [in Russian]

195 M. Asen-Palmer, K. Bartkowski, E. Gmelin et al., *Phys. Rev. B.* 56 (**1997**), p. 9431

196 K. Takyu, K. M. Itoh, K. Oka et al., *J. Appl. Phys.* 38 (**1999**), p. 1493

197 G. G. Devyatkin, A. V. Gusev, A. F. Khohlov et al., *Sov. Phys. Dokl.* 376(1) (**2001**), p. 62

198 B. F. Kane, *Nature* 393 (**1998**), p. 133

199 B. M. Andreev, Yu. A. Sakharnovsky, in *Isotopes: Properties, Production and Applications*, Ed. V. Yu. Baranov, IzdAT, Moscow, **2000**, p. 167 [in Russian]

200 Yu. D. Shipilov, A. V. Bespalov, *Proceedings of the 6th International Scientific Conference Physical and Chemical Processes on Selection of Atoms and Molecules*, TsNIIatominform, Moscow, Zvenigorod, **2001**, pp. 57–58

201 A. K. Kalitieevsky, O. N. Godisov, V. P. Liseykin et al., *Proceedings of the 6th International Scientific Conference Physical and Chemical Processes on Selection of Atoms and Molecules*, TsNIIatominform, Moscow, Zvenigorod, **2001**, pp. 84–86

202 J. F. Ziegler, H. W. Curtis, H. P. Muhlfeld et al., *IBM J. Res. Develop.* 40 (**1996**), p. 3

203 *Spectrum Sciences. News and Notes Quarterly* 1 (**1998**), p. 1

204 http://www.puretechnologies.com/

205 K. N. Tu, A. M. Gusak, M. Li, *J. Appl. Phys.* 93 (**2003**), p. 1335

206 G. L. Khorasanov, A. P. Ivanov, A. I. Blokhin, *Proceedings of the 6th International Scientific Conference Physical and Chemical Processes on Selection of Atoms and Molecules*, TsNIIatominform, Moscow, Zvenigorod, **2001**, pp. 22–28

207 A. L. Bortnyanskii, P. A. Bokhan, D. E. Zakrevsky et al., *New Industrial Technologies* 1 (**2003**), p. 33 [in Russian]

208 A. F. Chabak, A. S. Polevoi, in *Isotopes: Properties, Production and Applications*, Ed. V. Yu. Baranov, IzdAT, Moscow, **2000**, p. 496 [in Russian]

209 P. A. Bokhan, V. F. Gerasimov, V. I. Solomonov et al., *Sov. J. Quantum Electron.* 8 (**1978**), p. 1220

210 A. N. Soldatov, V. F. Fedorov, N. A. Yudin, *Sov. J. Quantum Electron.* 24 (**1994**), p. 677

211 P. A. Bokhan, B. A. Gerasimov, *Sov. J. Quantum Electron.* 9 (**1979**), p. 273

212 W. T. Walter, N. Solimene, M. Piltch et al., *IEEE J. Quantum Electron.* 2 (**1966**), p. 474

213 P. A. Bokhan, V. I. Silantiev, V. I. Solomonov, *Sov. J. Quantum Electron.* 10 (**1980**), p. 724

214 A. N. Maltsev, *Ph.D. thesis* Tomsk State University, (**1983**) [in Russian]

215 B. L. Borovich, N. I. Yurchenko, *Sov. J. Quantum Electron.* 13 (**1984**), p. 1386

216 X. Tiejun, Y. Zhixin, W. Yongjiang et al., *Acta Opt. Sin.* 5 (**1985**), p. 1104

217 B. L. Borovich, E. I. Molodykh, L. A. Ryazanskaya et al., *Sov. J. Quantum Electron.* 20 (**1990**), p. 1173

218 M. J. Kushner, B. E. Warner, *J. Appl. Phys.* 54 (**1983**), p. 2970

219 W. T. Walter, N. Solimene, G. M. Kull, in *Proc. Int. Conf. Laser 80*, STS Press, MacLin, VA **1981**, p. 148

220 A. A. Isaev, V. T. Mihkelsoo, G. G. Petrash et al., *Sov. J. Quantum Electron.* 18 (**1988**), p. 1577

221 R. J. Carman, M. J. Withford, D. J. W. Brown et al., *Opt. Commun.* 157 (**1998**), p. 99

222 P. A. Bokhan, D. E. Zakrevsky, *Tech. Phys.* 42(5) (**1997**), p. 504

223 P. A. Bokhan, D. E. Zakrevsky, *Quantum Electron.* 14 (**2002**), p. 602

224 P. A. Bokhan, E. I. Molodukh, in *Pulsed Metal Vapour Lasers*, Vol. 5, Kluwer, Dordrecht, **1996**, p. 137

225 P. A. Bokhan, *Sov. J. Quantum Electron.* 16 (**1986**), p. 1207

226 N. K. Vuchov, D. N. Astadjov, N. V. Sabotinov, *IEEE J. Quant. Electr.* QE-30 (**1994**), p. 750

227 P. B. Blau, *J. Appl. Phys.* 77 (**1995**), p. 2273

228 C. E. Webb, G. P. Hogan, *Pulsed Metal Vapour Lasers*, Vol. 5, Kluwer, Dordrecht, **1996**, p. 29

229 P. A. Bokhan, *Doctoral thesis* Tomsk State University, Tomsk, (**1988**) [in Russian]

230 K. Hayashi, E. Noda, Y. Iseki et al., *Proc. SPIE Int. Soc. Opt. Eng.* 1628 (**1992**), p. 44

231 D. J. W. Brown, R. Kunnemeyer, A. I. McIntosh, *IEEE J. Quant. Electr.* QE-26 (**1990**), p. 1609

232 A. V. Elesky, Yu. K. Zemsov, A. V. Rodin et al., *Sov. Phys. Dokl.* 20(1) (**1975**), p. 42

233 V. F. Elaev, A. N. Soldatov, G. B. Sukhanova, *High Temp.* 16 (**1980**), p. 1090 [in Russian]

234 V. D. Burlakov, A. N. Soldatov, G. M. Gorbunova, *Izv. Vyssh. Uchebn. Zaved., Fiz.*, Deposited Paper 2856-841 (**1984**) [in Russian]

235 R. J. Carman, D. J. W. Brown, I. A. Piper, *IEEE J. Quant. Electr.* QE-30 (**1994**), p. 1876

236 I. Smilanski, *Pulsed Metal Vapour Lasers*, Vol. 5, Kluwer, Dordrecht, **1996**, p. 87

237 R. J. Carman, *Pulsed Metal Vapour Lasers*, Vol. 5, Kluwer, Dordrecht, **1996**, p. 203

238 D. A. Leonard, *IEEE J. Quant. Electr.* QE-3 (**1967**), p. 380

239 V. M. Batenin, I. I. Klimovskii, M. A. Lesnoy et al., *Sov. J. Quantum Electron.* 7 (**1980**), p. 563

240 V. F. Elaev, A. N. Soldatov, G. B. Sukhanova, *High Temp.* 19 (**1981**), p. 426 [in Russian]

241 L. G. Dyachkov, G. R. Kobzev, *Sov. Tech. Phys.* 42(4) (**1997**), p. 346

242 V. M. Batenin, V. A. Burmakin, V. A. Vokhmin et al., *Sov. J. Quantum Electron.* 4 (**1977**), p. 891

243 P. A. Bokhan, *Sov. J. Quantum Electron.* 15 (**1985**), p. 622

244 P. A. Bokhan, *Sov. J. Quantum Electron.* 14 (**1986**), p. 839

245 V. V. Kazakov, C. V. Markova, G. G. Petrash, *Sov. J. Quantum Electron.* 13 (**1983**), p. 488

246 A. A. Isaev, V. V. Kazakov, M. A. Lesnoy et al., *Sov. J. Quantum Electron.* 16 (**1986**), p. 1517

247 B. R. Belostotsky, Yu. V. Lyubavsky, B. M. Ovchinnicov, *Fundamental Laser Technique*, Soviet Radio, Moscow, **1977** [in Russian]

248 V. V. Buchanov, E. I. Molodykh, N. I. Yurchenko, *Sov. J. Quantum Electron.* 13 (**1983**), p. 1022

249 P. A. Bokhan, V. I. Silantiev, V. I. Solomonov, *Certificate of Authorship 810032* **1979**

250 B. L. Borovich, E. P. Nalegach, V. M. Rybin, *Sov. J. Quantum Electron.* 14 (**1984**), p. 839

251 D. J. W. Brown, M. J. Withford, I. A. Piper, *IEEE J. Quant. Electr.* QE-37 (**2001**), p. 518

252 W. T. Walter, *Bull. Amer. Soc. Phys. Soc.* 1 (**1967**), p. 90

253 E. L. Latuch, L. M. Bukshpun, M. F. Sem, *Sov. J. Quantum Electron.* 18 (**1988**), p. 1098

254 P. A. Bokhan, D. E. Zakrevsky, *JETP Lett.* 62(1) (**1995**), p. 27

255 P. A. Bokhan, D. E. Zakrevsky, *Sov. Tech. Phys.* 42(4) (**1997**), p. 346

256 I. Smilanski, G. Erez, A. Kerman et al., *Opt. Commun.* 30 (**1979**), p. 70

257 O. R. Marasov, St. Stoilov, *Opt.Commun.* 46 (**1983**), p. 221

258 V. A. Burmakin, A. N. Evttyunin, M. A. Lesnoi, *Sov. J. Quantum Electron.* 9 (**1979**), p. 939

259 Z. G. Huang, K. Namba, F. Shimizu, *Jpn. J. Appl. Phys.* 25 (**1986**), p. 1677

260 A. D. Chursin, M. A. Kazaryan, I. S. Kolokolov et al., *Laser Physics* 7 (**2002**)

261 B. P. Yatsenko, A. M. Yudin, S. A. Motovilov et al., *Proceedings of the 6th International Scientific Conference Physical and Chemical Processes on Selection of Atoms and Molecules*, TsNIIatominform, Moscow, Zvenigorod, **2001**, pp. 116–122

262 N. A. Yudin, *Quantum Electron.* 30 (**2000**), p. 583

263 N. A. Yudin, *Quantum Electron.* 32 (**2002**), p. 815

264 K. I. Zemskov, A. A. Isaev, G. G. Petrash, *Quantum Electron.* 29 (**1999**), p. 462

265 K. Lark-Horovitz, in *Proceedings of the Conference on Semi-Conductinng Materials*, Ed. H. K. Hehish, Butterworth, London, **1951**, p. 47

266 *Neutron Transmutation Doping in Semiconductors*, Ed. J. Meese, Plenum, New York, London, **1979**

267 *Semiconductor Doping in Nuclear Reaction*, Ed. L. S. Smirnov, Nauka, Novosibirsk **1981** [in Russian]

268 I. S. Shlimak, *Fiz. Tv. Tela* 41(5) (**1999**), pp. 794–798

269 A. N. Ionov, M. N. Matveev, D. V. Smikk, *Zh. Tekhn. Fiz* 59 (**1989**), p. 169 [in Russian]

270 J. W. Coebett, G. D. Warkins, *Radiation Effects in Semiconductors* Plenum, New York, **1971**

Subject Index

AVLIS 1, 3, 4

cascade
– photoionization 18
– superluminescence 9, 106
coherent
– excitation 10
– interaction 6, 99, 109
– two-photon excitation 10, 14, 91
collimation of atomic beam 11
collisional quenching 46

density matrix 6, 9, 92, 104
detuning 91, 94, 109
deviation 91, 98, 99, 109, 142
Doppler
– profile 60
– width 60
driving generator 21, 26, 28–32, 141, 145

evaporation of material 10
exchange reaction 8, 55, 56

frequency
– detuning 15, 71, 73, 94, 98, 99
– doubling 18, 30, 35

harpoon model 41
hyperfine splitting 11, 62, 97

incoherent interaction 6, 7
ion extraction 10, 11

longitudinal gas circulation 64, 70, 71
Lorentz profile 60, 62, 75–77

oncoming beam 92, 109
optimization of excitation 163, 166

photochemical isotope separation 49, 51
photochemical reaction 17, 39, 40, 42, 45, 53, 57, 81, 87, 102
photoionization 9, 10, 12, 13, 15, 25, 51, 91, 93
plasma accelerators 12
pumping lasers 21, 32

repeated photoionization 95
resonance transfer 15, 70
Rydberg atom 51–53, 55

selectivity 40, 46, 52
single-photon excitation 14

transversal circulation 71, 75
tunable laser 4, 25, 35, 42, 59, 85, 95
two-photon excitation 14, 60, 91, 96, 98, 101, 103, 105, 112, 114